Learning Geospatial Analysis with Python

Second Edition

An effective guide to geographic information system and remote sensing analysis using Python 3

Joel Lawhead

BIRMINGHAM - MUMBAI

Learning Geospatial Analysis with Python
Second Edition

First published: October 2013

Second edition: December 2015

Production reference: 1211215

Published by Packt Publishing Ltd.
Livery Place
35 Livery Street
Birmingham B3 2PB, UK.

ISBN 978-1-78355-242-9

www.packtpub.com

Credits

Author
Joel Lawhead

Reviewers
Mark Cederholm
Truc Viet Le
John Maurer
Julia Wood

Commissioning Editor
Kartikey Pandey

Acquisition Editors
Kevin Colaco
Usha Iyer
Kartikey Pandey

Content Development Editor
Anish Sukumaran

Technical Editor
Manthan Raja

Copy Editor
Tasneem Fatehi

Project Coordinator
Izzat Contractor

Proofreader
Safis Editing

Indexer
Mariammal Chettiyar

Graphics
Jason Monteiro

Production Coordinator
Arvindkumar Gupta

Cover Work
Arvindkumar Gupta

About the Author

Joel Lawhead is a project management institute-certified Project Management Professional (PMP), certified GIS Professional (GISP), and the Chief Information Officer (CIO) of NVision Solutions Inc., an award-winning firm that specializes in geospatial technology integration and sensor engineering.

Joel began using Python in 1997 and started combining it with geospatial software development in 2000. He is the author of the first edition of *Learning Geospatial Analysis with Python* and *QGIS Python Programming Cookbook*, both by Packt Publishing. His Python cookbook recipes were featured in two editions of *Python Cookbook, O'Reilly Media*. He is also the developer of the widely-used, open source Python Shapefile Library (PyShp). He maintains the geospatial technical blog `http://geospatialpython.com/` and the Twitter feed, `@SpatialPython`, which discusses the use of the Python programming language in the geospatial industry.

In 2011, Joel reverse-engineered and published the undocumented shapefile spatial indexing format and assisted fellow geospatial Python developer, Marc Pfister, in reversing the algorithm used, allowing developers around the world to create better-integrated and more robust geospatial applications.

Joel serves as the lead architect, project manager, and co-developer for geospatial applications used by U.S. government agencies, including NASA, FEMA, NOAA, the U.S. Navy, and many other commercial and non-profit organizations. In 2002, he received the international Esri Special Achievement in GIS award for his work on the Real-Time Emergency Action Coordination Tool (REACT), for emergency management using geospatial analysis.

About the Reviewers

Mark Cederholm, GISP, has over 20 years of experience in developing GIS applications using various Esri technologies, from ARC/INFO AML to ArcObjects to ArcGIS Runtime and Web SDKs. He lives in Flagstaff, Arizona.

He has been a technical reviewer for the book, *Developing Mobile Web ArcGIS Applications*, *Packt Publishing*.

Truc Viet Le is currently a PhD candidate in information systems at the Singapore Management University. His research interests primarily involve novel methods for the modeling and predicting of human mobility patterns and trajectories, learning smart strategies for urban transportation, and traffic flow prediction from fine-grained GPS and sensor network data. He uses R and Python every day for his work, where he finds R superb for data manipulation/visualization and Python an ideal environment for machine learning tasks. He is also interested in applying data science for the social work and international development work. When not behind the computer screen, he is an avid traveler, adventurer, and an aspiring travel writer and photographer. His work portfolio and some of his writings can be found on his personal website at http://vietletruc.com/.

He spent a wonderful year at Carnegie Mellon University in Pittsburgh, Pennsylvania, while pursuing his PhD. Previously, he obtained his bachelor's and master's degrees from Nanyang Technological University in computer engineering and mathematical sciences.

John Maurer is a programmer and data manager at the Pacific Islands Ocean Observing System (PacIOOS) in Honolulu, Hawaii. He creates and configures web interfaces and data services to provide access, visualization, and mapping of oceanographic data from a variety of sources, including satellite remote sensing, forecast models, GIS layers, and in situ observations (buoys, sensors, shark tracking, and so on) throughout the insular Pacific. He obtained a graduate certificate in remote sensing as well as a master's degree in geography from the University of Colorado, Boulder, where he developed software to analyze ground-penetrating radar (GPR) for snow accumulation measurements on the Greenland ice sheet. While in Boulder, he worked with the National Snow and Ice Data Center (NSIDC) for eight years, sparking his initial interest in Earth science and all things geospatial: an unexpected but comfortable detour from his undergraduate degree in music, science, and technology at Stanford University.

Julia Wood is currently a Geospatial Information Sciences (GIS) analyst who spends her professional time completing projects as a contractor in the Washington D.C. area. She graduated magna cum laude from the University of Mary Washington in Fredericksburg, Virginia, in the spring of 2014 with a bachelor's degree in both history and geography as well as a minor in GIS. Though her career is still in its early stages, Julia has aspirations to keep growing her skill set, and working on this review has certainly helped expand her professional experience; she hopes to continue learning and eventually work toward a master's degree while still working full time. In her non-work life, she enjoys reading, crafting, cooking, and exploring the big city, one local restaurant at a time.

Reviewing this book for Packt Publishing was Julia's first professional reviewing experience and she hopes that she can pursue similar endeavors in the future.

I'd like to thank my parents, John and Diana, for always encouraging me to do well in my educational and professional endeavors; my sister, Sarrina, and my brother, Jonathan, for offering support and advice when I needed it; and my boyfriend, Max, and my cat, Coco, for keeping me company while I conducted the reviews for this book. A thank you to Packt for letting me be a part of this experience!

www.PacktPub.com

Support files, eBooks, discount offers, and more

For support files and downloads related to your book, please visit www.PacktPub.com.

Did you know that Packt offers eBook versions of every book published, with PDF and ePub files available? You can upgrade to the eBook version at www.PacktPub.com and as a print book customer, you are entitled to a discount on the eBook copy. Get in touch with us at service@packtpub.com for more details.

At www.PacktPub.com, you can also read a collection of free technical articles, sign up for a range of free newsletters and receive exclusive discounts and offers on Packt books and eBooks.

https://www2.packtpub.com/books/subscription/packtlib

Do you need instant solutions to your IT questions? PacktLib is Packt's online digital book library. Here, you can search, access, and read Packt's entire library of books.

Why subscribe?

- Fully searchable across every book published by Packt
- Copy and paste, print, and bookmark content
- On demand and accessible via a web browser

Free access for Packt account holders

If you have an account with Packt at www.PacktPub.com, you can use this to access PacktLib today and view 9 entirely free books. Simply use your login credentials for immediate access.

Table of Contents

Preface

The book starts with a background on geospatial analysis, offers a flow of techniques and technology used, and splits the field into its component specialty areas, such as Geographic Information Systems (GIS), remote sensing, elevation data, advanced modeling, and real-time data. The focus of the book is to lay a strong foundation in using the powerful Python language and framework in order to approach geospatial analysis effectively. In doing so, we'll focus on using pure Python as well as certain Python tools and APIs, and using generic algorithms. The readers will be able to analyze various forms of geospatial data that comes in and learn real-time data tracking and how to apply this to interesting scenarios.

While many third-party geospatial libraries are used throughout the examples, a special effort will made by us to use pure Python, with no dependencies, whenever possible. This focus on pure Python 3 examples is what will set this book apart from nearly all other information in this field. This book may be the only geospatial book using Python 3 on the market currently. We will also go through some popular libraries that weren't in the previous version of the book.

What this book covers

Chapter 1, *Learning Geospatial Analysis with Python*, introduces geospatial analysis as a way of answering questions about our world. The differences between GIS and remote sensing are explained. Common geospatial analysis processes are demonstrated using illustrations, basic formulas, pseudo code, and Python.

Chapter 2, *Geospatial Data*, explains the major categories of data as well as several newer formats that are becoming more and more common. Geospatial data comes in many forms. The most challenging part of geospatial analysis is acquiring the data that you need and preparing it for analysis. Familiarity with these data types is essential to understand geospatial analysis.

Chapter 3, *The Geospatial Technology Landscape*, tells you about the geospatial technology ecosystem that consists of thousands of software libraries and packages. This vast array of choices is overwhelming for newcomers to geospatial analysis. The secret to learning geospatial analysis quickly is understanding the handful of libraries and packages that really matter. Most other software is derived from these critical packages. Understanding the hierarchy of geospatial software and how it's used allows you to quickly comprehend and evaluate any geospatial tool.

Chapter 4, *Geospatial Python Toolbox*, introduces software and libraries that form the basis of the book and are used throughout. Python's role in the geospatial industry is elaborated: the GIS scripting language, mash-up glue language, and full-blown programming language. Code examples are used to teach data editing concepts, and many of the basic geospatial concepts in *Chapter 1*, *Learning Geospatial Analysis Using Python*, are also demonstrated in this chapter.

Chapter 5, *Python and Geographic Information Systems*, teaches you about simple yet practical Python GIS geospatial products using processes, which can be applied to a variety of problems.

Chapter 6, *Python and Remote Sensing*, shows you how to work with remote sensing geospatial data. Remote sensing includes some of the most complex and least-documented geospatial operations. This chapter will build a solid core for you and demystify remote sensing using Python.

Chapter 7, *Python and Elevation Data*, demonstrates the most common uses of elevation data and how to work with its unique properties. Elevation data deserves a chapter all on its own. Elevation data can be contained in almost any geospatial format but is used quite differently from other types of geospatial data.

Chapter 8, *Advanced Geospatial Python Modeling*, uses Python to teach you the true power of geospatial technology. Geospatial data editing and processing help us understand the world as it is. The true power of geospatial analysis is modeling. Geospatial models help us predict the future, narrow a vast field of choices down to the best options, and visualize concepts that cannot be directly observed in the natural world.

Chapter 9, *Real-Time Data*, examines the modern phenomena of geospatial analysis. A wise geospatial analyst once said, *"As soon as a map is created it is obsolete."* Until recently by the time you collected data about the earth, processed it, and created a geospatial product, the world it represented had already changed. But modern geospatial data shatter this notion. Data sets are available over the Internet which are up to the minute or even the second. This data fundamentally changes the way we perform geospatial analysis.

Chapter 10, Putting It All Together, combines the skills from the previous chapters step by step to build a generic corporate system to manage customer support requests and field support personnel that could be applied to virtually any organization.

What you need for this book

You will require Python (3.4 or higher), a minimum hardware requirement of a 300-MHz processor, 128 MB of RAM, 1.5 GB of available hard disk, and Windows, Linux, or OS X operating systems.

Who this book is for

This book is for anyone who wants to understand digital mapping and analysis and who uses Python or any other scripting language for the automation or crunching of data manually. This book primarily targets Python developers, researchers, and analysts who want to perform geospatial, modeling, and GIS analysis with Python.

Conventions

In this book, you will find a number of text styles that distinguish between different kinds of information. Here are some examples of these styles and an explanation of their meaning.

Code words in text, database table names, folder names, filenames, file extensions, pathnames, dummy URLs, user input, and Twitter handles are shown as follows: "The `pycsw` Python library implements the CSW standard."

A block of code is set as follows:

```
<?xml version="1.0" encoding="utf-8"?>
<kml xmlns="http://www.opengis.net/kml/2.2">
  <Placemark>
    <name>Mockingbird Cafe</name>
    <description>Coffee Shop</description>
    <Point>
      <coordinates>-89.329160,30.310964</coordinates>
    </Point>
  </Placemark>
</kml>
```

Any command-line input or output is written as follows:

```
t.pen(shown=False)
t.done()
```

 Warnings or important notes appear in a box like this.

 Tips and tricks appear like this.

Reader feedback

Feedback from our readers is always welcome. Let us know what you think about this book—what you liked or disliked. Reader feedback is important for us as it helps us develop titles that you will really get the most out of.

To send us general feedback, simply e-mail feedback@packtpub.com, and mention the book's title in the subject of your message.

If there is a topic that you have expertise in and you are interested in either writing or contributing to a book, see our author guide at www.packtpub.com/authors.

Customer support

Now that you are the proud owner of a Packt book, we have a number of things to help you to get the most from your purchase.

Downloading the example code

You can download the example code files from your account at http://www.packtpub.com for all the Packt Publishing books you have purchased. If you purchased this book elsewhere, you can visit http://www.packtpub.com/support and register to have the files e-mailed directly to you.

Downloading the color images of this book

We also provide you with a PDF file that has color images of the screenshots/diagrams used in this book. The color images will help you better understand the changes in the output. You can download this file from: `https://www.packtpub.com/sites/default/files/downloads/2429OS_ColorImages.pdf`.

Errata

Although we have taken every care to ensure the accuracy of our content, mistakes do happen. If you find a mistake in one of our books—maybe a mistake in the text or the code—we would be grateful if you could report this to us. By doing so, you can save other readers from frustration and help us improve subsequent versions of this book. If you find any errata, please report them by visiting `http://www.packtpub.com/submit-errata`, selecting your book, clicking on the **Errata Submission Form** link, and entering the details of your errata. Once your errata are verified, your submission will be accepted and the errata will be uploaded to our website or added to any list of existing errata under the Errata section of that title.

To view the previously submitted errata, go to `https://www.packtpub.com/books/content/support` and enter the name of the book in the search field. The required information will appear under the **Errata** section.

Piracy

Piracy of copyrighted material on the Internet is an ongoing problem across all media. At Packt, we take the protection of our copyright and licenses very seriously. If you come across any illegal copies of our works in any form on the Internet, please provide us with the location address or website name immediately so that we can pursue a remedy.

Please contact us at `copyright@packtpub.com` with a link to the suspected pirated material.

We appreciate your help in protecting our authors and our ability to bring you valuable content.

Questions

If you have a problem with any aspect of this book, you can contact us at `questions@packtpub.com`, and we will do our best to address the problem.

1
Learning Geospatial Analysis with Python

This chapter is an overview of geospatial analysis. We will see how geospatial technology is currently impacting our world with a case study of one of the worst disease epidemics that the world has ever seen and how geospatial analysis helped stop the deadly virus in its tracks. Next, we'll step through the history of geospatial analysis, which predates computers and even paper maps! Then, we'll examine why you might want to learn a programming language as a geospatial analyst as opposed to just using **geographic information system (GIS)** applications. We'll realize the importance of making geospatial analysis as accessible as possible to the broadest number of people. Then, we'll step through basic GIS and remote sensing concepts and terminology that will stay with you through the rest of the book. Finally, we'll use Python for geospatial analysis right in the first chapter by building the simplest possible GIS from scratch!

This book assumes some basic knowledge of Python, IT literacy, and at least an awareness of geospatial analysis. This chapter provides a foundation in geospatial analysis, which is needed to attack any subject in the areas of remote sensing and GIS, including the material in all the other chapters of the book.

Geospatial analysis and our world

On March 25, 2014, the world awoke to news from the United Nations **World Health Organization (WHO)** announcing the early stages of a deadly virus outbreak in West Africa. The fast-moving Ebola virus would spread rapidly over the coming summer months resulting in cases in six countries on three continents, including the United States and Europe.

Government and humanitarian agencies faced a vicious race against time to contain the outbreak. Patients without treatment died in as little as six days after symptoms appeared. The most critical piece of information was the location of new cases relative to the existing cases. The challenge they faced required reporting these new cases in mostly rural areas with limited infrastructure. Knowing where the virus existed in humans provided the foundation for all of the decisions that response agencies needed for containment. The location of cases defined the extent of the outbreak. It allowed governments to prioritize the distribution of containment resources and medical supplies. It allowed them to trace the disease to the first victim. It ultimately allowed them to see if they were making progress in slowing the disease.

This map is a relative heat map of the affected countries based on the number of cases documented and their location:

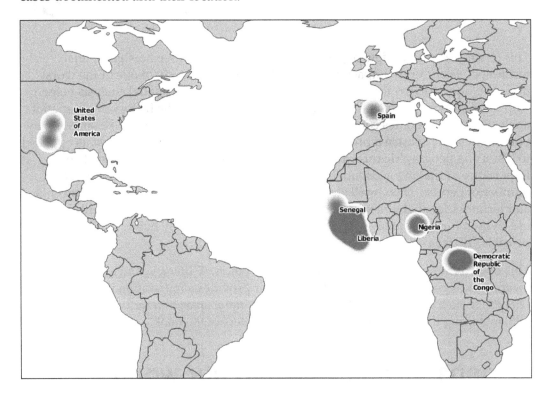

Unfortunately, the rural conditions and limited number of response personnel at the beginning of the outbreak resulted in a five-day reporting cycle to the Liberian Ministry of Health who initially tracked the virus. Authorities needed better information to bring the outbreak under control as new cases grew exponentially.

The solution came from a Liberian student using open source software from a non-profit Kenyan technology start-up called **Ushahidi**. Ushahidi is the Swahili word for *testimony* or *witness*. A team of developers in Kenya originally developed the system in 2008 to track reports of violence after the disputed presidential election there. Kpetermeni Siakor set up the system in Liberia in 2011 following a similarly controversial election. When the epidemic hit Liberia, Siakor turned Ushahidi into a disease-monitoring platform.

Siakor created a dispatch team of volunteers who received phone calls from the public reporting possible Ebola cases. The details were entered into the Ushahidi database, which was available on a web-based map almost instantly. The Liberian Ministry of Health and other humanitarian organizations could access the website, track the spread of the disease, and properly distribute supplies at health centers. This effort, amplified by the international response, would ultimately contain the epidemic globally. In 2015, cases are receding as the world monitors West African cases in anticipation of the last patient recovering. The following screenshot shows the latest Liberian public Ushahidi map as of April, 2015:

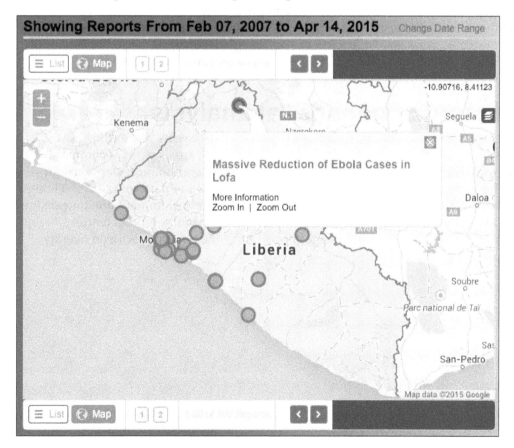

Relief workers also used the Ushahidi disaster mapping system to respond to the 2010 Haiti earthquake. Maps have always been used in disaster relief; however, the modern evolution of GPS-enabled mobile phones, web technology, and open source geospatial software have created a revolution in humanitarian efforts globally.

 The Ushahidi API has a Python library that you can find at
`https://github.com/ushahidi/ushapy`.

Beyond disasters

The application of geospatial modeling to disaster relief is one of the most recent and visible case studies. However, the use of geospatial analysis has been increasing steadily over the last 15 years. In 2004, the U.S. Department of Labor declared the geospatial industry as one of 13 high-growth industries in the United States expected to create millions of jobs in the coming decades.

Geospatial analysis can be found in almost every industry, including real estate, oil and gas, agriculture, defense, politics, health, transportation, and oceanography, to name a few. For a good overview of how geospatial analysis is used in dozens of different industries, visit `http://geospatialrevolution.psu.edu`.

History of geospatial analysis

Geospatial analysis can be traced as far back as 15,000 years ago to the Lascaux cave in southwestern France. In this cave, Paleolithic artists painted commonly hunted animals and what many experts believe are astronomical star maps for either religious ceremonies or potentially even migration patterns of prey. Though crude, these paintings demonstrate an ancient example of humans creating abstract models of the world around them and correlating spatial-temporal features to find relationships. The following image shows one of the paintings with an overlay illustrating the star maps:

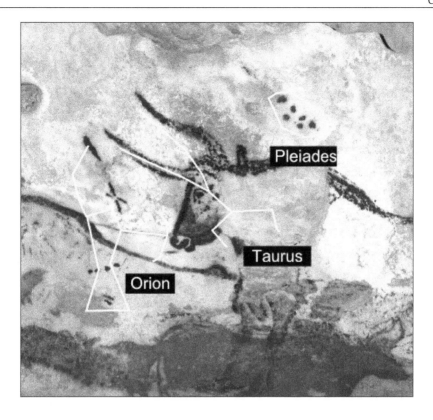

Over the centuries, the art of cartography and science of land surveying has developed, but it wasn't until the 1800s that significant advances in geographic analysis emerged. Deadly cholera outbreaks in Europe between 1830 and 1860 led geographers in Paris and London to use geographic analysis for epidemiological studies.

In 1832, Charles Picquet used different half-toned shades of gray to represent deaths per thousand citizens in the 48 districts of Paris as part of a report on the cholera outbreak. In 1854, Dr. John Snow expanded on this method by tracking a cholera outbreak in London as it occurred. By placing a point on a map of the city each time a fatality was diagnosed, he was able to analyze the clustering of cholera cases. Snow traced the disease to a single water pump and prevented further cases. The map has three layers with streets, an X for each pump, and dots for each cholera death:

Geospatial analysis wasn't just used for the war on diseases. For centuries, generals and historians have used maps to understand human warfare. A retired French engineer named Charles Minard produced some of the most sophisticated infographics ever drawn between 1850 and 1870. The term infographics is too generic to describe these drawings because they have strong geographic components. The quality and detail of these maps make them fantastic examples of geographic information analysis even by today's standards. Minard released his masterpiece, *Carte figurative des pertes successives en hommes de l'Armée Française dans la campagne de Russie 1812-1813*, in 1869, which depicted the decimation of Napoleon's army in the Russian campaign of 1812. The map shows the size and location of the army over time along with prevailing weather conditions. The following graphic contains four different series of information on a single theme. It is a fantastic example of geographic analysis using pen and paper. The size of the army is represented by the widths of the brown and black swaths at a ratio of one millimeter for every 10,000 men. The numbers are also written along the swaths. The brown-colored path shows soldiers who entered Russia, while the black represents the ones who made it out. The map scale is shown to the right in the center as one French league (2.75 miles or 4.4 kilometers). The chart at the bottom runs from right to left and depicts the brutal freezing temperatures experienced by the soldiers on the return march home from Russia:

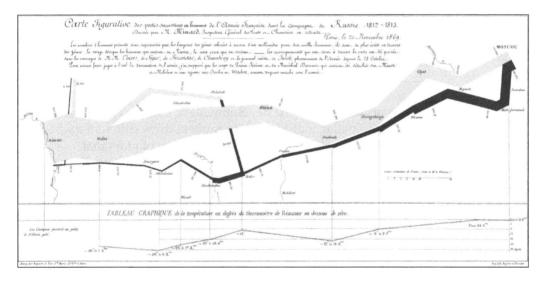

While far more mundane than a war campaign, Minard released another compelling map cataloguing the number of cattle sent to Paris from around France. Minard used pie charts of varying sizes in the regions of France to show each area's variety and volume of cattle that was shipped:

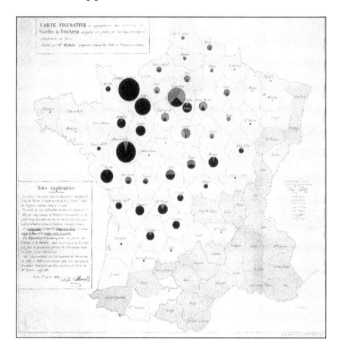

In the early 1900s, mass printing drove the development of the concept of map layers—a key feature of geospatial analysis. Cartographers drew different map elements (vegetation, roads, and elevation contours) on plates of glass that could then be stacked and photographed to be printed as a single image. If the cartographer made a mistake, only one plate of glass had to be changed instead of the entire map. Later, the development of plastic sheets made it even easier to create, edit, and store maps in this manner. However, the layering concept for maps as a benefit to analysis would not come into play until the modern computer age.

Geographic information systems

Computer mapping evolved with the computer itself in the 1960s. However, the origin of the term GIS began with the Canadian Department of Forestry and Rural Development. Dr. Roger Tomlinson headed a team of 40 developers in an agreement with IBM to build the **Canada Geographic Information System (CGIS)**. The CGIS tracked the natural resources of Canada and allowed the profiling of these features for further analysis. The CGIS stored each type of land cover as a different layer. It also stored data in a Canadian-specific coordinate system suitable for the entire country that was devised for optimal area calculations. While the technology used was primitive by today's standards, the system had phenomenal capability at that time. The CGIS included software features that seem quite modern: map projection switching, rubber sheeting of scanned images, map scale change, line smoothing and generalization to reduce the number of points in a feature, automatic gap closing for polygons, area measurement, dissolving and merging of polygons, geometric buffering, creation of new polygons, scanning, and digitizing of new features from the reference data.

The National Film Board of Canada produced a documentary in 1967 on the CGIS, which can be seen at the following URL:

```
http://video.esri.com/watch/128/data-for-decision_
comma_-1967-short-version
```

Tomlinson is often called the *father of GIS*. After launching the CGIS, he earned his doctorate from the University of London with his 1974 dissertation, entitled *The application of electronic computing methods and techniques to the storage, compilation, and assessment of mapped data*, which describes GIS and geospatial analysis. Tomlinson now runs his own global consulting firm, Tomlinson Associates Ltd., and remains an active participant in the industry. He is often found delivering the keynote addresses at geospatial conferences.

CGIS is the starting point of geospatial analysis as defined by this book. However, this book would not have been written if not for the work of Howard Fisher and the Harvard Laboratory for Computer Graphics and Spatial Analysis at the Harvard Graduate School of Design. His work on the SYMAP GIS software that outputs maps to a line printer, started an era of development at the laboratory, which produced two other important packages and, as a whole, permanently defined the geospatial industry. SYMAP led to other packages including **GRID** and the **Odyssey** project from the same lab. GRID was a raster-based GIS system that used cells to represent geographic features instead of geometry. GRID was written by Carl Steinitz and David Sinton. The system later became IMGRID. Next came Odyssey. Odyssey was a team effort led by Nick Chrisman and Denis White. It was a system of programs that included many advanced geospatial data management features that are typical of modern geodatabase systems. Harvard attempted to commercialize these packages with limited success. However, their impact is still seen today. Virtually, every existing commercial and open source package owes something to these code bases.

Howard Fisher produced a 1967 film using the output from SYMAP to show the urban expansion of Lansing, Michigan, from 1850 to 1965 by hand coding decades of property information into the system. The analysis took months but would take only a few minutes to create now using modern tools and data. You can see the film at `http://youtu.be/xj8DQ7IQ8_o`.

There are now dozens of **graphical user interface (GUI)** geospatial desktop applications available today from companies including Esri, ERDAS, Intergraph, and ENVI, to name a few. Esri is the oldest, continuously operating GIS software company, which started in the late 1960s. In the open source realm, packages including **Quantum GIS (QGIS)** and **Geographic Resources Analysis Support System (GRASS)** are widely used. Beyond comprehensive desktop software packages, software libraries for the building of new software exist in the thousands.

Remote sensing

Remote sensing is the collection of information about an object without making physical contact with that object. In the context of geospatial analysis, the object is usually the Earth. Remote sensing also includes the processing of the collected information. The potential of geographic information systems is limited only by the available geographic data. The cost of land surveying, even using a modern GPS, to populate a GIS has always been resource-intensive. The advent of remote sensing not only dramatically reduced this cost of geospatial analysis, but it took the field in entirely new directions. In addition to powerful reference data for GIS systems, remote sensing has made possible the automated and semi-automated generation of GIS data by extracting features from images and geographic data. The eccentric French photographer, Gaspard-Félix Tournachon, also known as *Nadar*, took the first aerial photograph in 1858 from a hot-air balloon over Paris:

The value of a true bird's-eye view of the world was immediately apparent. As early as 1920, books on aerial photo interpretation began to appear.

When the United States entered the Cold War with the Soviet Union after World War II, aerial photography to monitor military capability became prolific with the invention of the American **U-2** spy plane. The U-2 spy plane could fly at 70,000 feet, putting it out of the range of existing anti-aircraft weapons designed to reach only 50,000 feet. The American U-2 flights over Russia ended when the Soviets finally shot down a U-2 and captured the pilot.

However, aerial photography had little impact on modern geospatial analysis. Planes could only capture small footprints of an area. Photographs were tacked to walls or examined on light tables but not in the context of other information. Though extremely useful, aerial photo interpretation was simply another visual perspective.

The game changer came on October 4, 1957, when the Soviet Union launched the **Sputnik 1** satellite. The Soviets had scrapped a much more complex and sophisticated satellite prototype because of manufacturing difficulties. Once corrected, this prototype would later become Sputnik 3. Instead, they opted for a simple metal sphere with four antennae and a simple radio transmitter. Other countries including the United States were also working on satellites. The satellite initiatives were not entirely a secret. They were driven by scientific motives as part of the **International Geophysical Year (IGY)**. Advancement in rocket technology made artificial satellites a natural evolution for Earth science. However, in nearly every case, each country's defense agency was also heavily involved. Similar to the Soviets, other countries were struggling with complex satellite designs packed with scientific instruments. The Soviets' decision to switch to the simplest possible device for the sole reason of launching a satellite before the Americans was effective. Sputnik was visible in the sky as it passed over and its radio pulse could be heard by amateur radio operators. Despite Sputnik's simplicity, it provided valuable scientific information that could be derived from its orbital mechanics and radio frequency physics.

The Sputnik program's biggest impact was on the American space program. America's chief adversary had gained a tremendous advantage in the race to space. The United States ultimately responded with the Apollo moon landings. However, before this, the U.S. launched a program that would remain a national secret until 1995. The classified **CORONA** program resulted in the first pictures from space. The U.S. and Soviet Union had signed an agreement to end spy plane flights but satellites were conspicuously absent from the negotiations. The following map shows the CORONA process. Dashed lines are the satellite flight paths, the longer white tubes are the satellites, the smaller white cones are the film canisters, and the black blobs are the control stations that triggered the ejection of the film so that a plane could catch it in the sky:

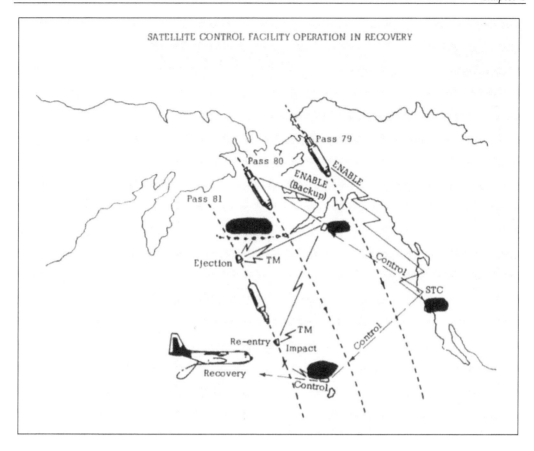

SATELLITE CONTROL FACILITY OPERATION IN RECOVERY

The first CORONA satellite was a four-year effort with many setbacks. However, the program ultimately succeeded. The difficulty of satellite imaging, even today, is retrieving the images from space. The CORONA satellites used canisters of black and white film that were ejected from the vehicle once exposed. As the film canister parachuted to Earth, a U.S. military plane would catch the package in midair. If the plane missed the canister, it would float for a brief duration in the water before sinking into the ocean to protect the sensitive information. The U.S. continued to develop the CORONA satellites until they matched the resolution and photographic quality of the U-2 spy plane photos. The primary disadvantages of the CORONA instruments were reusability and timeliness. Once out of film, a satellite could no longer be of service. Additionally, the film recovery was on a set schedule making the system unsuitable to monitor real-time situations. The overall success of the CORONA program, however, paved the way for the next wave of satellites, which ushered in the modern era of remote sensing.

Due to the CORONA program's secret status, its impact on remote sensing was indirect. Photographs of the Earth taken on manned U.S. space missions inspired the idea of a civilian-operated remote sensing satellite. The benefits of such a satellite were clear but the idea was still controversial. Government officials questioned whether a satellite was as cost-efficient as aerial photography. The military were worried that the public satellite could endanger the secrecy of the CORONA program. Other officials worried about the political consequences of imaging other countries without permission. However, the **Department of the Interior (DOI)** finally won permission for NASA to create a satellite to monitor Earth's surface resources.

On July 23, 1972, NASA launched the **Earth Resources Technology Satellite (ERTS)**. The ERTS was quickly renamed to **Landsat-1**. The platform contained two sensors. The first was the **Return Beam Vidicon (RBV)** sensor, which was essentially a video camera. It was built by the radio and television giant, **Radio Corporation of America (RCA)**. The RBV immediately had problems, which included disabling the satellite's altitude guidance system. The second attempt at a satellite was the highly experimental **Multispectral Scanner (MSS)**. The MSS performed flawlessly and produced superior results than the RBV. The MSS captured four separate images at four different wavelengths of the light reflected from the Earth's surface.

This sensor had several revolutionary capabilities. The first and most important capability was the first global imaging of the planet scanning every spot on the Earth every 16 days. The following image from the U.S. **National Aeronautics and Space Administration (NASA)** illustrates this flight and collection pattern that is a series of overlapping swaths as the sensor orbits the Earth capturing tiles of data each time the sensor images a location on the Earth:

It also recorded light beyond the visible spectrum. While it did capture green and red light visible to the human eye, it also scanned near-infrared light at two different wavelengths not visible to the human eye. The images were stored and transmitted digitally to three different ground stations in Maryland, California, and Alaska. The multispectral capability and digital format meant that the aerial view provided by Landsat wasn't just another photograph from the sky. It was beaming down data. This data could be processed by computers to output derivative information about the Earth in the same way a GIS provided derivative information about the Earth by analyzing one geographic feature in the context of another. NASA promoted the use of Landsat worldwide and made the data available at very affordable prices to anyone who asked.

This global imaging capability led to many scientific breakthroughs including the discovery of previously unknown geography as late as 1976. For example, using Landsat imagery, the Government of Canada located a tiny uncharted island inhabited by polar bears. They named the new landmass Landsat Island.

Landsat 1 was followed by six other missions and turned over to the **National Oceanic and Atmospheric Administration (NOAA)** as the responsible agency. Landsat 6 failed to achieve orbit due to a ruptured manifold, which disabled its maneuvering engines. During some of these missions, the satellites were managed by the **Earth Observation Satellite (EOSAT)** company, now called **Space Imaging**, but returned to government management by the Landsat 7 mission. The following image from NASA is a sample of a Landsat 7 product:

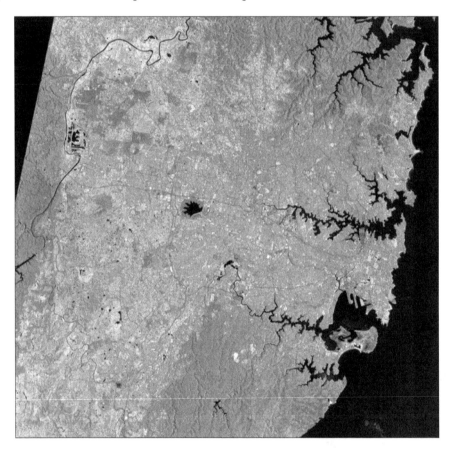

The **Landsat Data Continuity Mission (LDCM)** was launched on February 13, 2013, and began collecting images on April 27, 2013, as part of its calibration cycle to become Landsat 8. The LDCM is a joint mission between NASA and the **U. S. Geological Survey (USGS)**.

Elevation data

A **Digital Elevation Model (DEM)** is a three-dimensional representation of a planet's terrain. In the context of this book, this planet is the Earth. The history of digital elevation models is far less complicated than remotely-sensed imagery but no less significant. Before computers, representations of elevation data were limited to topographic maps created through traditional land surveys. Technology existed to create three-dimensional models from stereoscopic images or physical models from materials such as clay or wood, but these approaches were not widely used for geography.

The concept of digital elevation models began in 1986 when the French space agency, **Centre national d'études spatiales (CNES)**, launched its **SPOT-1** satellite, which included a stereoscopic radar. This system created the first usable DEM. Several other U.S. and European satellites followed this model with similar missions. In February, 2000, the **Space Shuttle Endeavour** conducted the **Shuttle Radar Topography Mission (SRTM)**, which collected elevation data over 80% of the Earth's surface using a special radar antenna configuration that allowed a single pass. This model was surpassed in 2009 by the joint U.S. and Japanese mission using the **Advanced Spaceborne Thermal Emission and Reflection Radiometer (ASTER)** sensor aboard NASA's Terra satellite. This system captured 99% of the Earth's surface but has proven to have minor data issues. As the Space Shuttle's orbit did not cross the Earth's poles, it did not capture the entire surface. SRTM remains the gold standard. The following image from the USGS shows a colorized DEM known as a hillshade. Greener areas are lower elevations while yellow and brown areas are mid-range to high elevations:

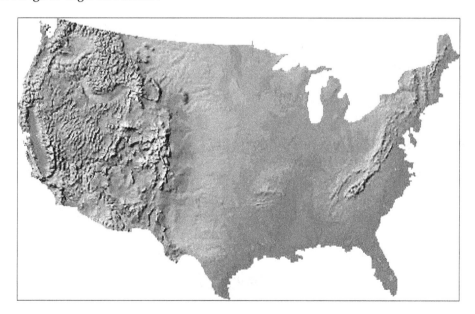

Recently, more ambitious attempts at a worldwide elevation dataset are underway in the form of **TerraSAR-X** and **TanDEM-X** satellites launched by Germany in 2007 and 2010, respectively. These two radar elevation satellites worked together to produce a global DEM called **WorldDEM** that was released on April 15, 2014. This dataset has a relative accuracy of two meters and an absolute accuracy of four meters.

Computer-aided drafting

Computer-aided drafting (CAD) is worth mentioning, though it does not directly relate to geospatial analysis. The history of CAD system development parallels and intertwines with the history of geospatial analysis. CAD is an engineering tool used to model two- and three-dimensional objects usually for engineering and manufacturing. The primary difference between a geospatial model and CAD model is that a geospatial model is referenced to the Earth, whereas a CAD model can possibly exist in abstract space. For example, a three-dimensional blueprint of a building in a CAD system would not have a latitude or longitude, but in a GIS, the same building model would have a location on the Earth. However, over the years, CAD systems have taken on many features of GIS systems and are commonly used for smaller GIS projects. Likewise, many GIS programs can import CAD data that has been georeferenced. Traditionally, CAD tools were designed primarily for the engineering of data that was not geospatial.

However, engineers who became involved with geospatial engineering projects, such as designing a city utility electric system, would use the CAD tools that they were familiar with in order to create maps. Over time, both the GIS software evolved to import the geospatial-oriented CAD data produced by engineers, and CAD tools evolved to support geospatial data creation and better compatibility with GIS software. **AutoCAD** by Autodesk and **ArcGIS** by Esri were the leading commercial packages to develop this capability and the **Geospatial Data Abstraction Library** (**GDAL**) **OGR** library developers added CAD support as well.

Geospatial analysis and computer programming

Modern geospatial analysis can be conducted with the click of a button in any of the easy-to-use commercial or open source geospatial packages. So then, why would you want to use a programming language to learn this field? The most important reasons are as follows:

- You want complete control of the underlying algorithms, data, and execution

- You want to automate specific, repetitive analysis tasks with minimal overhead from a large, multipurpose geospatial framework
- You want to create a program that's easy to share
- You want to learn geospatial analysis beyond pushing buttons in software

The geospatial industry is gradually moving away from the traditional workflow in which teams of analysts use expensive desktop software to produce geospatial products. Geospatial analysis is being pushed towards automated processes that reside in the cloud. End user software is moving towards task-specific tools, many of which are accessed from mobile devices. Knowledge of geospatial concepts and data as well as the ability to build custom geospatial processes is where the geospatial work in the near future lies.

Object-oriented programming for geospatial analysis

Object-oriented programming is a software development paradigm in which concepts are modeled as objects that have properties and behaviors represented as attributes and methods, respectively. The goal of this paradigm is more modular software in which one object can inherit from one or more other objects to encourage software reuse.

The Python programming language is known for its ability to serve multiple roles as a well-designed, object-oriented language, a procedural scripting language, or even a functional programming language. However, you never completely abandon object-oriented programming in Python because even its native data types are objects and all Python libraries, known as modules, adhere to a basic object structure and behavior.

Geospatial analysis is the perfect activity for object-oriented programming. In most object-oriented programming projects, the objects are abstract concepts such as database connections that have no real-world analogy. However, in geospatial analysis, the concepts modeled are, well, real-world objects! The domain of geospatial analysis is the Earth and everything on it. Trees, buildings, rivers, and people are all examples of objects within a geospatial system.

A common example in literature for newcomers to object-oriented programming is the concrete analogy of a cat. Books on object-oriented programming frequently use some form of the following example.

Imagine that you are looking at a cat. We know some information about the cat, such as its name, age, color, and size. These features are the properties of the cat. The cat also exhibits behaviors such as eating, sleeping, jumping, and purring. In object-oriented programming, objects have properties and behaviors too. You can model a real-world object such as the cat in our example or something more abstract such as a bank account.

Most concepts in object-oriented programming are far more abstract than the simple cat paradigm or even the bank account. However, in geospatial analysis, the objects that are modeled remain concrete, such as the simple cat analogy, and in many cases are living, breathing cats. Geospatial analysis allows you to continue with the simple cat analogy and even visualize it. The following map represents the feral cat population of Australia using data provided by the **Atlas of Living Australia (ALA)**:

Importance of geospatial analysis

Geospatial analysis helps people make better decisions. It doesn't make the decision for you, but it can answer critical questions that are at the heart of the choice to be made and often cannot be answered any other way. Until recently, geospatial technology and data were tools available only to governments and well-funded researchers. However, in the last decade, data has become much more widely available and software much more accessible to anyone.

In addition to freely available government satellite imagery, many local governments now conduct aerial photo surveys and make the data available online. The ubiquitous Google Earth provides a cross-platform spinning globe view of the Earth with satellite and aerial data, streets, points of interest, photographs, and much more. Google Earth users can create custom **Keyhole Markup Language** (KML) files, which are XML files to load and style data to the globe. This program, and similar tools, are often called geographic exploration tools because they are excellent data viewers but provide very limited data analysis capability.

The ambitious **OpenStreetMap** project (`http://www.openstreetmap.org`) is a crowd-sourced, worldwide, geographic basemap containing most layers commonly found in a GIS. Nearly every mobile phone now contains a GPS along with mobile apps to collect GPS tracks as points, lines, or polygons. Most phones will also tag photos taken with the phone's camera with GPS coordinates. In short, anyone can be a geospatial analyst.

The global population has reached seven billion people. The world is changing faster than ever before. The planet is undergoing environmental changes never seen before in recorded history. Faster communication and transportation increase the interaction between us and the environment in which we live. Managing people and resources safely and responsibly is more challenging than ever. Geospatial analysis is the best approach to understanding our world more efficiently and deeply. The more politicians, activists, relief workers, parents, teachers, first responders, medical professionals, and small businesses harness the power of geospatial analysis, the more our potential for a better, healthier, safer, and fairer world will be realized.

Geographic information system concepts

In order to begin geospatial analysis, it is important to understand some key underlying concepts unique to the field. The list isn't long, but nearly every aspect of analysis traces back to one of these ideas.

Thematic maps

As its name suggests, a thematic map portrays a specific theme. A general reference map visually represents features as they relate geographically for navigation or planning. A thematic map goes beyond location to provide the geographic context for information around a central idea. Usually, a thematic map is designed for a targeted audience to answer specific questions. The value of thematic maps lies in what they do not show. A thematic map will use minimal geographic features to avoid distracting the reader from the theme. Most thematic maps include political boundaries such as country or state borders but omit navigational features, such as street names or points of interest beyond major landmarks that orient the reader. The cholera map by Dr. John Snow earlier in this chapter is a perfect example of a thematic map. Common uses for thematic maps are visualizing health issues, such as disease, election results, and environmental phenomena such as rainfall. These maps are also the most common output of geospatial analysis. The following map from the United States Census Bureau shows cancer mortality rates by state:

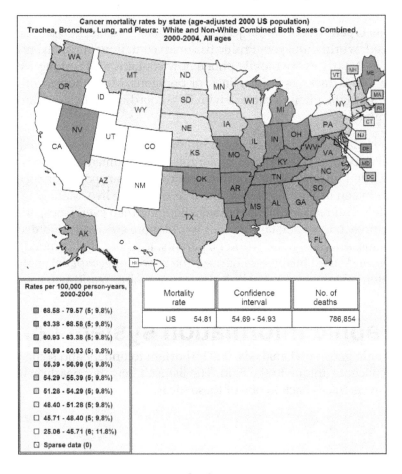

Thematic maps tell a story and are very useful. However, it is important to remember that while thematic maps are models of reality as any other map, they are also generalizations of information. Two different analysts using the same source of information will often come up with very different thematic maps depending on how they analyze and summarize the data. They may also choose to focus on different aspects of the dataset. The technical nature of thematic maps often leads people to treat them as if they are scientific evidence. However, geospatial analysis is often inconclusive. While the analysis may be based on scientific data, the analyst does not always follow the rigor of the scientific method. In his classic book, *How to Lie with Maps, Mark Monmonier, University of Chicago Press*, demonstrates in great detail how maps are easily manipulated models of reality, which are commonly abused. This fact doesn't degrade the value of these tools. The legendary statistician, *George Box*, wrote in his 1987 book, *Empirical Model-Building and Response Surfaces* that, *"Essentially, all models are wrong, but some are useful."* Thematic maps have been used as guides to start (and end) wars, stop deadly diseases in their tracks, win elections, feed nations, fight poverty, protect endangered species, and rescue those impacted by disaster. Thematic maps may be the most useful models ever created.

Spatial databases

In its purest form, a database is simply an organized collection of information. A **database management system (DBMS)** is an interactive suite of software that can interact with a database. People often use the word *database* as a catch-all term referring to both the DBMS and underlying data structure. Databases typically contain alphanumeric data and, in some cases, binary large objects or blobs, which can store binary data such as images. Most databases also allow a relational database structure in which entries in normalized tables can be referenced to each other in order to create many-to-one and one-to-many relationships among data.

Spatial databases, also known as geodatabases, use specialized software to extend a traditional **relational database management system (RDBMS)** to store and query data defined in two-dimensional or three-dimensional space. Some systems also account for a series of data over time. In a spatial database, attributes about geographic features are stored and queried as traditional relational database structures. The spatial extensions allow you to query geometries using **Structured Query Language (SQL)** in a similar way as traditional database queries. Spatial queries and attribute queries can also be combined to select results based on both location and attributes.

Spatial indexing

Spatial indexing is a process that organizes the geospatial vector data for faster retrieval. It is a way of prefiltering the data for common queries or rendering. Indexing is commonly used in large databases to speed up returns to queries. Spatial data is no different. Even a moderately-sized geodatabase can contain millions of points or objects. If you perform a spatial query, every point in the database must be considered by the system in order to include it or eliminate it in the results. Spatial indexing groups data in ways that allow large portions of the dataset to be eliminated from consideration by doing computationally simpler checks before going into a detailed and slower analysis of the remaining items.

Metadata

Metadata is defined as data about data. Accordingly, geospatial metadata is data about geospatial datasets that provide traceability for the source and history of a dataset as well as summary of the technical details. Metadata also provides long-term preservation of information holdings. Geospatial metadata can be represented by several possible standards. One of the most prominent standards is the international standard, **ISO 19115-1**, which includes hundreds of potential fields to describe a single geospatial dataset. Additionally, the **ISO 19115-2** includes extensions for geospatial imagery and gridded data. Some example fields include spatial representation, temporal extent, and lineage. The primary use of metadata is cataloging datasets. Modern metadata can be ingested by geographic search engines making it potentially discoverable by other systems automatically. It also lists points of contact for a dataset if you have questions. Metadata is an important support tool for geospatial analysts and adds credibility and accessibility to your work. The **Open Geospatial Consortium** (**OGC**) created the **Catalog Service for the Web** (**CSW**) to manage metadata. The `pycsw` Python library implements the CSW standard. You can learn more about it at `http://pycsw.org`.

Map projections

Map projections have entire books devoted to them and can be a challenge for new analysts. If you take any three-dimensional object and flatten it on a plane, such as your screen or a sheet of paper, the object is distorted. Many grade school geography classes demonstrate this concept by having students peel an orange and then attempt to lay the peel flat on their desk in order to understand the resulting distortion. The same effect occurs when you take the round shape of the Earth and project it on a computer screen.

In geospatial analysis, you can manipulate this distortion to preserve common properties, such as area, scale, bearing, distance, or shape. There is no one-size-fits-all solution to map projections. The choice of projection is always a compromise of gaining accuracy in one dimension in exchange for error in another. Projections are typically represented as a set of over 40 parameters as either XML or a text format called **Well-Known Text** (**WKT**), which is used to define the transformation algorithm.

The **International Association of Oil & Gas Producers** (**IOGP**) maintains a registry of the most known projections. The organization was formerly known as the **European Petroleum Survey Group** (**EPSG**). The entries in the registry are still known as EPSG codes. The EPSG maintained the registry as a common benefit for the oil and gas industry, which is a prolific user of geospatial analysis for energy exploration. At the last count, this registry contained over 5,000 entries.

As recently as 10 years ago, map projections were a primary concern for a geospatial analyst. Data storage was expensive, high-speed Internet was rare, and cloud computing didn't really exist. Geospatial data was typically exchanged among small groups working in separate areas of interest. The technology constraints at the time meant that geospatial analysis was highly localized. Analysts would use the best projection for their area of interest. Data in different projections cannot be displayed on the same map because they represent two different models of the Earth. Any time an analyst received data from a third party, it had to be reprojected before using it with the existing data. This process was tedious and time-consuming. Most geospatial data formats do not provide a way to store the projection information. This information is stored in an ancillary file as text or XML usually. As analysts didn't exchange data often, many people wouldn't bother defining projection information. Every analyst's nightmare was to come across an extremely valuable dataset missing the projection information. It rendered the dataset useless. The coordinates in the file are just numbers and offer no clue to the projection. With over 5,000 choices, it was nearly impossible to guess.

Now, thanks to modern software and the Internet making data exchange easier and more common, nearly every data format has added a metadata format that defines the projection or places it in the file header if supported. Advances in technology have also allowed for global basemaps, which allow for more common uses of projections such as the common Google Mercator projection used for Google Maps. This projection is also known as **Web Mercator** and uses code EPSG:3857 (or the deprecated EPSG:900913). Geospatial portal projects such as `OpenStreetMap.org` and `NationalAtlas.gov` have consolidated datasets for much of the world in common projections. Modern geospatial software can also reproject data on the fly saving the analyst the trouble of preprocessing the data before using it. Closely related to map projections are geodetic datums. A datum is a model of the Earth's surface used to match the location of features on the Earth to a coordinate system. One common datum is called WGS 84 that is used by GPS devices.

Rendering

The exciting part of geospatial analysis is visualization. As geospatial analysis is a computer-based process, it is good to be aware of how geographic data appears on a computer screen.

Geographic data including points, lines, and polygons are stored numerically as one or more points, which come in (x,y) pairs or (x,y,z) tuples. The x represents the horizontal axis on a graph. The y represents the vertical axis. The z represents terrain elevation. In computer graphics, a computer screen is represented by an x and y axis. A z axis is not used because the computer screen is treated as a two-dimensional plane by most graphics software APIs. However, as desktop computing power continues to improve, three-dimensional maps are starting to become more common.

Another important factor is screen coordinates versus world coordinates. Geographic data is stored in a coordinate system representing a grid overlaid on the Earth, which is three-dimensional and round. Screen coordinates, also known as pixel coordinates, represent a grid of pixels on a flat, two-dimensional computer screen. Mapping x and y world coordinates to pixel coordinates is fairly straightforward and involves a simple scaling algorithm. However, if a z coordinate exists, then a more complicated transform must be performed to map coordinates from three-dimensional space to a two-dimensional plane. These transformations can be computationally costly and therefore slow if not handled correctly.

In the case of remote sensing data, the challenge is typically the file size. Even a moderately-sized satellite image that is compressed can be tens, if not hundreds, of megabytes. Images can be compressed using lossless or lossy methods. Lossless methods use tricks to reduce the file size without discarding any data. Lossy compression algorithms reduce the file size by reducing the amount of data in the image while avoiding a significant change in the appearance of the image. Rendering an image on the screen can be computationally-intensive. Most remote sensing file formats allow for the storing of multiple lower-resolution versions of the image — called overviews or pyramids — for the sole purpose of faster rendering at different scales. When zoomed out from the image to a scale where you couldn't see the detail of the full resolution image, a preprocessed, lower-resolution version of the image is displayed quickly and seamlessly.

Remote sensing concepts

Most of the GIS concepts described also apply to raster data. However, raster data has some unique properties as well. Earlier in this chapter, in the history of remote sensing, the focus was on Earth imaging from aerial platforms. It is important to note that raster data can come in many forms including ground-based radar, laser range finders, and other specialized devices to detect gases, radiation, and other forms of energy in a geographic context. For the purpose of this book, we will focus on remote sensing platforms that capture large amounts of Earth data. These sources included Earth imaging systems, certain types of elevation data, and some weather systems where applicable.

Images as data

Raster data is captured digitally as square tiles. This means that the data is stored on a computer as a numerical array of rows and columns. If the data is multispectral, the dataset will usually contain multiple arrays of the same size, which are geospatially referenced together to represent a single area on the Earth. These different arrays are called bands. Any numerical array can be represented on a computer as an image. In fact, all computer data is ultimately numbers. It is important in geospatial analysis to think of images as a numeric array because mathematical formulas are used to process them.

In remotely sensed images, each pixel represents both space (location on the Earth of a certain size) and the reflectance captured as light reflected from the Earth at this location into space. So, each pixel has a ground size and contains a number representing the intensity. As each pixel is a number, we can perform mathematical equations on this data to combine data from different bands and highlight specific classes of objects in the image. If the wavelength value is beyond the visible spectrum, we can highlight features not visible to the human eye. Substances such as chlorophyll in plants can be greatly contrasted using a specific formula called **Normalized Difference Vegetation Index (NDVI)**.

By processing remotely sensed images, we can turn this data into visual information. Using the NDVI formula, we can answer the question, what is the relative health of the plants in this image? You can also create new types of digital information, which can be used as input for computer programs to output other types of information.

Remote sensing and color

Computer screens display images as combinations of **Red, Green, and Blue** (**RGB**) to match the capability of the human eye. Satellites and other remote sensing imaging devices can capture light beyond this visible spectrum. On a computer, wavelengths beyond the visible spectrum are represented in the visible spectrum so that we can see them. These images are known as **false color** images. In remote sensing, for instance, infrared light makes moisture highly visible. This phenomenon has a variety of uses such as monitoring ground saturation during a flood or finding hidden leaks in a roof or levee.

Common vector GIS concepts

This section will discuss the different types of GIS processes commonly used in geospatial analysis. This list is not exhaustive; however, it provides you with the essential operations that all other operations are based on. If you understand these operations, you can quickly understand much more complex processes as they are either derivatives or combinations of these processes.

Data structures

GIS vector data uses coordinates consisting of, at a minimum, an x horizontal value and a y vertical value to represent a location on the Earth. In many cases, a point may also contain a z value. Other ancillary values are possible including measurements or timestamps.

These coordinates are used to form points, lines, and polygons to model real-world objects. Points can be a geometric feature in and of themselves or they can connect line segments. Closed areas created by line segments are considered polygons. Polygons model objects such as buildings, terrain, or political boundaries.

A GIS feature can consist of a single point, line, or polygon or it can consist of more than one shape. For example, in a GIS polygon dataset containing world country boundaries, the Philippines, which is made up of 7,107 islands, would be represented as a single country made up of thousands of polygons.

Vector data typically represents topographic features better than raster data. Vector data has better accuracy potential and is more precise. However, to collect vector data on a large scale is also traditionally more costly than raster data.

Two other important terms related to vector data structures are **bounding box** and **convex hull**. The bounding box or minimum bounding box is the smallest possible square that contains all of the points in a dataset. The following image demonstrates a bounding box for a collection of points:

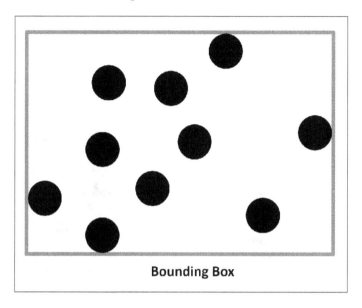

Bounding Box

The convex hull of a dataset is similar to the bounding box, but instead of a square, it is the smallest possible polygon that can contain a dataset. The bounding box of a dataset always contains its convex hull. The following image shows the same point data as the previous example with the convex hull polygon shown in red:

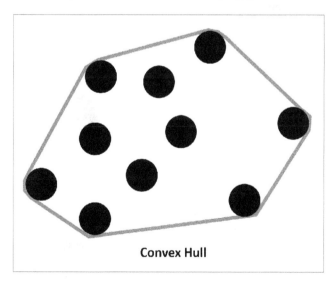

Convex Hull

Buffer

A buffer operation can be applied to spatial objects including points, lines, or polygons. This operation creates a polygon around the object at a specified distance. Buffer operations are used for proximity analysis, for example, establishing a safety zone around a dangerous area. In the following image, the black shapes represent the original geometry while the red outlines represent the larger buffer polygons generated from the original shape:

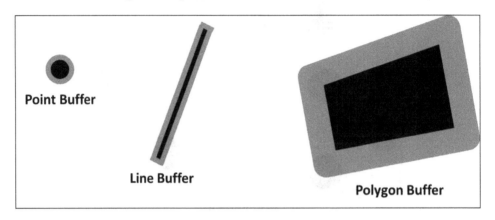

Dissolve

A dissolve operation creates a single polygon out of adjacent polygons. A common use for a dissolve operation is to merge two adjacent properties in a tax database that have been purchased by a single owner. Dissolves are also used to simplify data extracted from remote sensing:

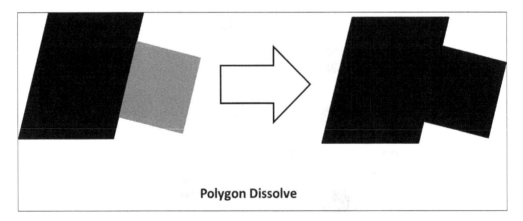

Generalize

Objects that have more points than necessary for the geospatial model can be generalized to reduce the number of points used to represent the shape. This operation usually requires a few attempts to get the optimal number of points without compromising the overall shape. It is a data optimization technique to simplify data for the efficiency of computing or better visualization. This technique is useful in web-mapping applications. Computer screens have a resolution of 72 **dots per inch** (**dpi**). Highly-detailed point data, which would not be visible, can be reduced so that less bandwidth is used to send a visually equivalent map to the user:

Polygon Generalization

Intersection

An intersection operation is used to see if one part of a feature intersects with one or more features. This operation is for spatial queries in proximity analysis and is often a follow-on operation to a buffer analysis:

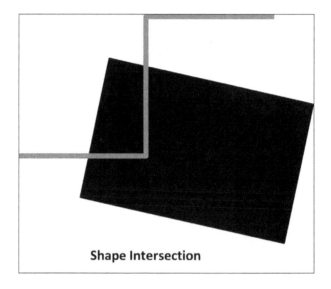

Shape Intersection

Merge

A merge operation combines two or more non-overlapping shapes in a single multishape object. Multishape objects mean that the shapes maintain separate geometries but are treated as a single feature with a single set of attributes by the GIS:

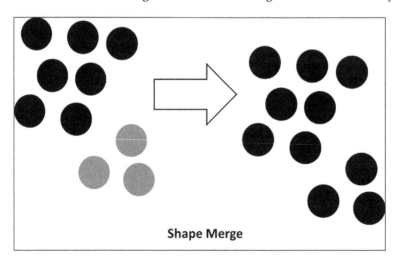

Shape Merge

Point in polygon

A fundamental geospatial operation is checking to see whether a point is inside a polygon. This one operation is the atomic building block of many different types of spatial queries. If the point is on the boundary of the polygon, it is considered inside. Very few spatial queries exist that do not rely on this calculation in some way. However, it can be very slow on a large number of points.

The most common and efficient algorithm to detect if a point is inside a polygon is called the **ray casting algorithm**. First, a test is performed to see if the point is on the polygon boundary. Next, the algorithm draws a line from the point in question in a single direction. The program counts the number of times the line crosses the polygon boundary until it reaches the bounding box of the polygon. The bounding box is the smallest box that can be drawn around the entire polygon. If the number is odd, the point is inside. If the number of boundary intersections is even, the point is outside:

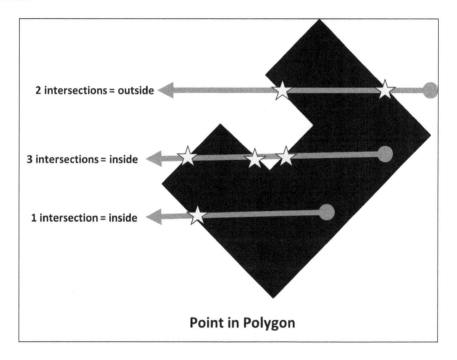

Point in Polygon

Union

The union operation is less common but very useful to combine two or more overlapping polygons in a single shape. It is similar to dissolve, but in this case, the polygons are overlapping as opposed to being adjacent. Usually, this operation is used to clean up automatically-generated feature datasets from remote sensing operations:

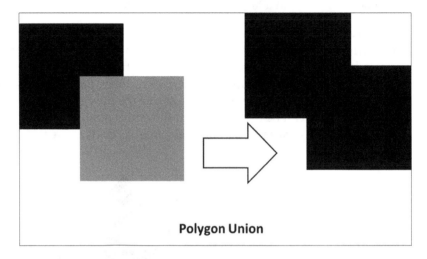

Polygon Union

Join

A join or SQL join is a database operation used to combine two or more tables of information. Relational databases are designed to avoid storing redundant information for one-to-many relationships. For example, a U.S. state may contain many cities. Rather than creating a table for each state containing all of its cities, a table of states with numeric IDs is created, while a table for all the cities in every state is created with a state numeric ID. In a GIS, you can also have spatial joins that are part of the spatial extension software for a database. In spatial joins, combine the attributes to two features in the same way that you do in a SQL join, but the relation is based on the spatial proximity of the two features. To follow the previous cities example, we could add the county name that each city resides in using a spatial join. The cities layer could be loaded over a county polygon layer whose attributes contain the county name. The spatial join would determine which city is in which county and perform a SQL join to add the county name to each city's attribute row.

Geospatial rules about polygons

In geospatial analysis, there are several general rules of thumb regarding polygons that are different from mathematical descriptions of polygons:

- Polygons must have at least four points—the first and last points must be the same

- A polygon boundary should not overlap itself

- Polygons in a layer shouldn't overlap

- A polygon in a layer inside another polygon is considered as a hole in the underlying polygon

Different geospatial software packages and libraries handle exceptions to these rules differently and can lead to confusing errors or software behaviors. The safest route is to make sure that your polygons obey these rules. There is one more important piece of information about polygons. A polygon is by definition a closed shape, which means that the first and last vertices of the polygon are identical. Some geospatial software will throw an error if you haven't explicitly duplicated the first point as the last point in the polygon dataset. Other software will automatically close the polygon without complaining. The data format that you use to store your geospatial data may also dictate how polygons are defined. This issue is a gray area and so it didn't make the polygon rules, but knowing this quirk will come in handy someday when you run into an error that you can't explain easily.

Common raster data concepts

As mentioned earlier, remotely-sensed raster data is a matrix of numbers. Remote sensing contains thousands of operations that can be performed on data. This field changes on almost a daily basis as new satellites are put into space and computer power increases. Despite its decades-long history, we haven't even scratched the surface of the knowledge that this field can provide to the human race. Once again, similar to the common GIS processes, this minimal list of operations gives you the basis to evaluate any technique used in remote sensing.

Band math

Band math is multidimensional array mathematics. In array math, arrays are treated as single units, which are added, subtracted, multiplied, and divided. However, in an array, the corresponding numbers in each row and column across multiple arrays are computed simultaneously. These arrays are termed matrices and computations involving matrices are the focus of linear algebra.

Change detection

Change detection is the process of taking two images of the same location at different times and highlighting the changes. A change can be due to the addition of something on the ground, such as a new building or the loss of a feature such as coastal erosion. There are many algorithms to detect changes among images and also determine qualitative factors such as how long ago the change took place. The following image from a research project by the U.S. **Oak Ridge National Laboratory** (**ORNL**) shows rainforest deforestation between 1984 and 2000 in the state of Rondonia, Brazil. Colors are used to show how recently the forest was cut. Green represents virgin rainforests, white is a forest cut within two years of the end of the date range, red is within 22 years, and the other colors fall in between as described in the legend:

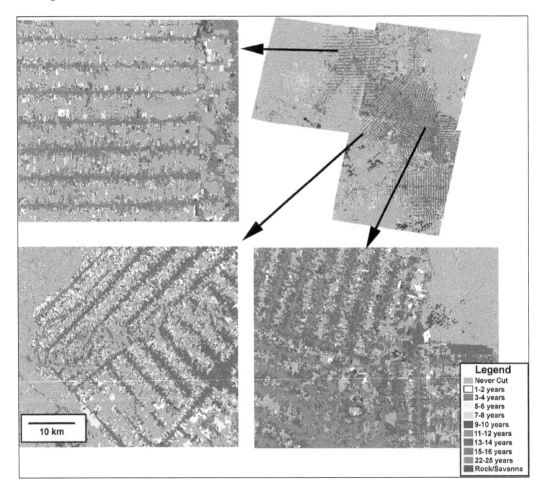

Legend
Never Cut
1-2 years
3-4 years
5-6 years
7-8 years
9-10 years
11-12 years
13-14 years
15-16 years
22-25 years
Rock/Savanna

10 km

Histogram

A histogram is the statistical distribution of values in a dataset. The horizontal axis represents a unique value in a dataset while the vertical axis represents the frequency of this unique value in the raster. A histogram is a key operation in most raster processing. It can be used for everything from enhancing contrast in an image to serving as a basis for object classification and image comparison. The following example from NASA shows a histogram for a satellite image that has been classified into different categories representing the underlying surface features:

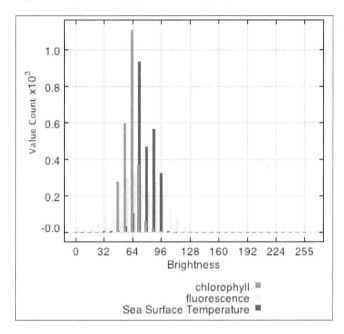

Feature extraction

Feature extraction is the manual or automatic digitization of features in an image to points, lines, or polygons. This process serves as the basis for the vectorization of images in which a raster is converted to a vector dataset. An example of feature extraction is extracting a coastline from a satellite image and saving it as a vector dataset. If this extraction is performed over several years, you could monitor the erosion or other changes along this coastline.

Supervised classification

Objects on the Earth reflect different wavelengths of light depending on the materials that they are made of. In remote sensing, analysts collect wavelength signatures for specific types of land cover (for example, concrete) and build a library for a specific area. A computer can then use this library to automatically locate classes in the library in a new image of the same area.

Unsupervised classification

In an unsupervised classification, a computer groups pixels with similar reflectance values in an image without any other reference information other than the histogram of the image.

Creating the simplest possible Python GIS

Now that we have a better understanding of geospatial analysis, the next step is to build a simple GIS using Python called SimpleGIS. This small program will be a technically complete GIS with a geographic data model and the ability to render the data as a visual thematic map showing the population of different cities.

The data model will also be structured so that you can perform basic queries. Our SimpleGIS will contain the state of Colorado, three cities, and population counts for each city.

Most importantly, we will demonstrate the power and simplicity of Python programming by building this tiny system in pure Python. We will only use modules available in the standard Python distribution without downloading any third-party libraries.

Getting started with Python

As stated earlier, this book assumes that you have some basic knowledge of Python. The examples in this book are based on Python 3.4.3, which you can download here:

https://www.python.org/downloads/release/python-343/

The only module used in the following example is the `turtle` module that provides a very simple graphics engine based on the `Tkinter` library included with Python. If you used the installers for Windows or Mac OS X, the `Tkinter` library should be included already. If you compiled Python yourself or are using a distribution from somewhere besides `Python.org`, then make sure that you can import the `turtle` module by typing the following at a command prompt to run the `turtle` demo script:

```
python -m turtle
```

If your Python distribution does not have Tkinter, you can find information on installing it from the following page. The information is for Python 2.3 but the process is still the same:

`http://tkinter.unpythonic.net/wiki/How_to_install_Tkinter`

The official Python wiki page for Tkinter can be found here:

`https://wiki.python.org/moin/TkInter`

The documentation for Tkinter is in the Python Standard Library documentation that can be found at `https://docs.python.org/2/library/tkinter.html`.

If you are new to Python, *Dive into Python* is a free online book, which covers all the basics of Python and will bring you up to speed. For more information, refer to `http://www.diveintopython.net`.

Building SimpleGIS

The code is divided into two different sections. The first is the data model section and the second is the map renderer that draws this data. For the data model, we will use simple Python lists. A Python list is a native data type, which serves as a container for other Python objects in a specified order. Python lists can contain other lists and are great for simple data structures. They also map well to more complex structures or even databases if you decide you want to develop your script further.

The second portion of the code will render the map using the Python turtle graphics engine. We will have only one function in the GIS that converts the world coordinates, in this case, longitude and latitude, into pixel coordinates. All graphics engines have an origin point of (0,0) that is usually in the top-left or lower-left corner of the canvas. Turtle graphics are designed to teach programming visually. The turtle graphics canvas uses an origin of **(0,0)** in the center, similar to a graphing calculator. The following image illustrates the type of Cartesian graph that the turtle module uses. In the following graph, some points are plotted in both positive and negative space:

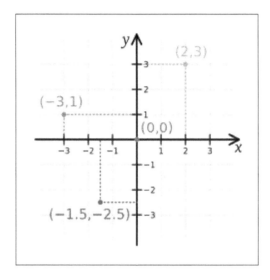

This also means that the turtle graphics engine can have negative pixel coordinates, which is uncommon for graphics canvases. However, for this example, the turtle module is the quickest and simplest way to render our map.

Step by step

You can run this program interactively in the Python interpreter or you can save the complete program as a script and run it. The Python interpreter is an incredibly powerful way to learn new concepts because it gives you real-time feedback on errors or unexpected program behavior. You can easily recover from these issues and try something else until you get the results that you want:

1. In Python, you usually import modules at the beginning of the script so we'll import the `turtle` module first. We'll use Python's import as feature to assign the module the name `t` to save space and time when typing turtle commands:

    ```
    import turtle as t
    ```

2. Next, we'll set up the data model starting with some simple variables that allow us to access list indexes by name instead of numbers to make the code easier to follow. Python lists index the contained objects starting with the number 0. So, if we want to access the first item in a list called `myList`, we would reference it as follows:

```
myList[0]
```

3. To make our code easier to read, we can also use a variable name assigned to commonly used indexes:

```
firstItem = 0
```

```
myList[firstItem]
```

Downloading the example code

You can download the example code files for all the Packt books that you have purchased from your account at http://www.packtpub.com. If you purchased this book elsewhere, you can visit http://www.packtpub.com/support and register in order to have the files e-mailed to you directly.

In computer science, assigning commonly used numbers to an easy-to-remember variable is a common practice. These variables are called constants.

So, for our example, we'll assign constants for some common elements used for all the cities. All cities will have a name, one or more points, and a population count:

```
NAME = 0
POINTS = 1
POP = 2
```

4. Now, we'll set up the data for Colorado as a list with name, polygon points, and population. Note that the coordinates are a list within a list:

```
state = ["COLORADO", [[-109, 37],[-109, 41],[-102, 41],[-102,
37]], 5187582]
```

5. The cities will be stored as nested lists. Each city's location consists of a single point as a longitude and latitude pair. These entries will complete our GIS data model. We'll start with an empty list called cities and then append the data to this list for each city:

```
cities = []
cities.append(["DENVER",[-104.98, 39.74], 634265])
cities.append(["BOULDER",[-105.27, 40.02], 98889])
cities.append(["DURANGO",[-107.88,37.28], 17069])
```

6. We will now render our GIS data as a map by first defining a map size. The width and height can be anything that you want depending on your screen resolution:

```
map_width = 400
map_height = 300
```

7. In order to scale the map to the graphics canvas, we must first determine the bounding box of the largest layer, which is the state. We'll set the map's bounding box to a global scale and reduce it to the size of the state. To do so, we'll loop through the longitude and latitude of each point and compare it with the current minimum and maximum x and y values. If it is larger than the current maximum or smaller than the current minimum, we'll make this value the new maximum or minimum, respectively:

```
minx = 180
maxx = -180
miny = 90
maxy = -90
forx,y in state[POINTS]:
if x < minx: minx = x
  elif x > maxx: maxx = x
  if y < miny: miny = y
  elif y > maxy: maxy = y
```

8. The second step to scaling is to calculate a ratio between the actual state and the tiny canvas that we will render it on. This ratio is used for coordinate to pixel conversion. We get the size along the x and y axes of the state and then we divide the map width and height by these numbers to get our scaling ratio:

```
dist_x = maxx - minx
dist_y = maxy - miny
x_ratio = map_width / dist_x
y_ratio = map_height / dist_y
```

9. The following function, called convert(), is our only function in SimpleGIS. It transforms a point in the map coordinates from one of our data layers to pixel coordinates using the previous calculations. You'll notice that, at the end, we divide the map width and height in half and subtract it from the final conversion to account for the unusual center origin of the turtle graphics canvas. Every geospatial program has some form of this function:

```
def convert(point):
  lon = point[0]
```

```
lat = point[1]
x = map_width - ((maxx - lon) * x_ratio)
y = map_height - ((maxy - lat) * y_ratio)
# Python turtle graphics start in the
# middle of the screen
# so we must offset the points so they are centered
x = x - (map_width/2)
y = y - (map_height/2)
return [x,y]
```

10. Now for the exciting part! We're ready to render our GIS as a thematic map. The `turtle` module uses the concept of a cursor called a pen. Moving the cursor around the canvas is exactly the same as moving a pen around a piece of paper. The cursor will draw a line when you move it. So, you'll notice that throughout the code, we use the `t.up()` and `t.down()` commands to pick the pen up when we want to move to a new location and put it down when we're ready to draw. We have some important steps in this section. As the border of Colorado is a polygon, we must draw a line between the last point and first point to close the polygon. We can also leave out the closing step and just add a duplicate point to the Colorado dataset. Once we draw the state, we'll use the `write()` method to label the polygon:

```
t.up()
first_pixel = None
for point in state[POINTS]:
  pixel = convert(point)
  if not first_pixel:
    first_pixel = pixel
  t.goto(pixel)
  t.down()
t.goto(first_pixel)
t.up()
t.goto([0,0])
t.write(state[NAME], align="center", font=("Arial",16,"bold"))
```

11. If we were to run the code at this point, we would see a simplified map of the state of Colorado, as shown in the following screenshot:

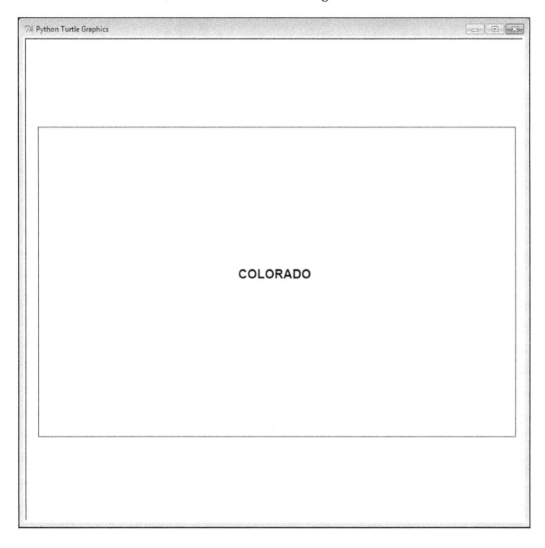

 If you do try to run the code, you'll need to temporarily add the following line at the end, or the Tkinter window will close as soon as it finishes drawing:

```
t.done()
```

12. Now, we'll render the cities as point locations and label them with their names and population. As the cities are a group of features in a list, we'll loop through them to render them. Instead of drawing lines by moving the pen around, we'll use the turtle `dot()` method to plot a small circle at the pixel coordinate returned by our `SimpleGISconvert()` function. We'll then label the dot with the city's name and add the population. You'll notice that we must convert the population number to a string in order to use it in the turtle `write()` method. To do so, we use Python's built-in function, `str()`:

```
#for city in cities:
pixel = convert(city[POINTS])
t.up()
t.goto(pixel)
# Place a point for the city
t.dot(10)
# Label the city
t.write(city[NAME] + ", Pop.: " + str(city[POP]), align="left")
t.up()
```

13. Now we will perform one last operation to prove that we have created a real GIS. We will perform an attribute query on our data to determine which city has the largest population. Then, we'll perform a spatial query to see which city lies the furthest west. Finally, we'll print the answers to our questions on our thematic map page safely out of the range of the map.

14. As our query engine, we'll use Python's built-in `min()` and `max()` functions. These functions take a list as an argument and return the minimum and maximum values of this list. These functions have a special feature called a key argument that allows you to sort complex objects. As we are dealing with nested lists in our data model, we'll take advantage of the key argument in these functions. The key argument accepts a function that temporarily alters the list for evaluation before a final value is returned. In this case, we want to isolate the population values for comparison and then the points. We could write a whole new function to return the specified value, but we can use Python's lambda keyword instead. The lambda keyword defines an anonymous function that is used inline. Other Python functions can be used inline, for example, the string function, `str()`, but they are not anonymous. This temporary function will isolate our value of interest.

15. So our first question is, which city has the largest population?

```
biggest_city = max(cities, key=lambda city:city[POP])
t.goto(0,-200)
t.write("The biggest city is: " +  biggest_city[NAME])
```

16. The next question is, which city lies the furthest west?

```
western_city = min(cities, key=lambda city:city[POINTS])
t.goto(0,-220)
t.write("The western-most city is: " + western_city[NAME])
```

17. In the preceding query, we use Python's built in `min()` function to select the smallest longitude value and this works because we represented our city locations as longitude and latitude pairs. It is possible to use different representations for points including possible representations where this code would need modification to work correctly. However, for our `SimpleGIS`, we are using a common point representation to make it as intuitive as possible.

These last two commands are just for clean up purposes. First, we hide the cursor. Then, we call the turtle `done()` method, which will keep the turtle graphics window with our map open until we choose to close it using the close handle at the top of the window.

```
t.pen(shown=False)
t.done()
```

Whether you followed along using the Python interpreter or you ran the complete program as a script, you should see the following map rendered in real time:

Congratulations! You have followed in the footsteps of Paleolithic hunters, the *father of GIS* Dr. Roger Tomlinson, geospatial pioneer Howard Fisher, and game-changing humanitarian programmers to create a functional, extensible, and technically complete geographic information system. It took less than 60 lines of pure Python code! You will be hard-pressed to find a programming language that can create a complete GIS using only its core libraries in such a finite amount of readable code as Python. Even if you did, it is highly unlikely that the language would survive the geospatial Python journey that you'll take through the rest of this book.

As you can see, there is lots of room for expansion of `SimpleGIS`. Here are some other ways that you might expand this simple tool using the reference material for Tkinter and Python linked at the beginning of this section:

- Create an overview map in the top-right corner with a U.S. border outline and Colorado's location in the U.S.
- Add color for visual appeal and further clarity
- Create a map key for different features
- Make a list of states and cities and add more states and cities
- Add a title to the map
- Create a bar chart to compare population numbers visually

The possibilities are endless. `SimpleGIS` can also be used as a way to quickly test and visualize geospatial algorithms that you come across. If you want to add more data layers, you can create more lists, but these lists would become difficult to manage. In this case, you can use another Python module included in the standard distribution. The `SQLite` module provides a SQL-like database in Python that can be saved to disk or run in memory.

Summary

Well done! You are now a geospatial analyst. In this chapter, you learned the state of the art of geospatial analysis through the story of Ushahidi and the Ebola virus. You learned the history of geospatial analysis and the technologies that support it. You became familiar with foundational GIS and remote sensing concepts that will serve you through the rest of the book. You actually built a working GIS that can be expanded to do whatever you can imagine! In the next chapter, we'll tackle the data formats that you'll encounter as geospatial analysts. Geospatial analysts spend far more time dealing with data than actually performing analysis. Understanding the data that you're working with is essential to working efficiently and having fun.

2
Geospatial Data

One of the most challenging aspects of geospatial analysis is the data. Geospatial data already includes dozens of file formats and database structures and continues to evolve and grow to include new types of data and standards. Additionally, almost any file format can technically contain geospatial information simply by adding a location. In this chapter, we'll examine some common traits of geospatial data. Then we'll look at some of the most widely used vector data types followed by raster data types. We'll gain some insight into newer, more complex types including point cloud data and web services.

An overview of common data formats

As a geospatial analyst, you may frequently encounter the following general data types:

- Spreadsheets and comma-separated files (CSV files) or tab-separated files (TSV files)
- Geotagged photos
- Lightweight binary points, lines, and polygons
- Multi-gigabyte satellite or aerial images
- Elevation data such as grids, point clouds, or integer-based images
- XML files
- JSON files
- Databases (both servers and file databases)
- Web services

Each format contains its own challenges for access and processing. When you perform analysis on data, usually you have to do some form of preprocessing first. You might clip or subset a satellite image of a large area down to just your area of interest, or you might reduce the number of points in a collection to just the ones meeting certain criteria in your data model. A good example of this type of preprocessing is the SimpleGIS example at the end of *Chapter 1, Learning Geospatial Analysis with Python*. The state dataset included just the state of Colorado rather than all the 50 states. The city dataset included only three sample cities demonstrating three levels of population along with different relative locations.

The common geospatial operations in Chapter 1 are the building blocks for this type of preprocessing. However, it is important to note that there has been a gradual shift in the field of geospatial analysis towards readily available base maps. Until around 2004, geospatial data was difficult to acquire and desktop computing power was much less than it is today. Preprocessing data was an absolute first step for any geospatial project. However, in 2004, Google released Google Maps not long after Google Earth. Microsoft had also been developing a technology acquisition called **TerraServer**, which they relaunched around this time. In 2004, the **Open Geospatial Consortium** (OGC) updated the version of its **Web Map Service (WMS)** to 1.3.0. This same year, Esri also released version 9 of their ArcGIS server system. These innovations were driven by Google's web map tiling model that allowed for smooth, global, scrolling maps at many different resolutions often called slippy maps.

People used map servers on the Internet before Google Maps, most famously with the MapQuest driving directions website. However, these map servers offered only small amounts of data at a time and usually over limited areas. The Google tiling system converted global maps to tiered image tiles for both images and mapping data. These were served dynamically using JavaScript and the browser-based XMLHttpRequest API, more commonly known as **asynchronous JavaScript and XML (AJAX)**. Google's system scaled to millions of users using ordinary web browsers. More importantly, it allowed programmers to leverage JavaScript programming to create mash-ups to use the Google Maps JavaScript API to add additional data to the maps. The mash-up concept is actually a distributed geospatial layers system. Users can combine and recombine data from different locations into a single map as long as the data is web accessible. Other commercial and open source systems quickly mimicked the idea of distributed layers.

Notable examples of distributed geospatial layers are **OpenLayers**, which provide an open source Google-like API that has now gone beyond Google's API offering additional features. Complimentary to OpenLayers is **OpenStreetMap**, which is the open source answer to the tiled map services consumed by systems like OpenLayers. OpenStreetMap has global, street-level vector data and other spatial features collected from available government data sources and the contributions of thousands of editors worldwide. OpenStreetMap's data maintenance model is similar to the way the Wikipedia online encyclopedia crowd-sources information creation and update for articles. Recently, even more mapping APIs have appeared, including Leaflet and Mapbox, which continue to increase in flexibility, simplicity, and capability.

The mash-up revolution had interesting and beneficial side effects on data. Geospatial data is traditionally difficult to obtain. The cost of collecting, processing, and distributing data kept geospatial analysis constrained to those who could afford this steep overhead cost by producing data or purchasing it. For decades, geospatial analysis was the tool of governments, very large organizations, and universities. Once the web mapping trend shifted to large-scale, globally-tiled maps, organizations began essentially providing base map layers for free in order to draw developers to their platform. The massively scalable global map system required massively scalable, high-resolution data to be useful. Geospatial software producers and data providers wanted to maintain their market share and kept up with the technology trend.

Geospatial analysts benefited greatly from this market shift in several ways. First of all, data providers began distributing data in a common projection called Mercator. The Mercator projection is a nautical navigation projection introduced over 400 years ago. As mentioned in *Chapter 1, Learning Geospatial Analysis with Python*, all projections have practical benefits as well as distortions. The distortion in the Mercator projection is size. In a global view, Greenland appears bigger than the continent of South America. However, like every projection, it also has a benefit. Mercator preserves angles. Predictable angles allowed medieval navigators to draw straight bearing lines when plotting a course across oceans. Google Maps didn't launch with Mercator. However, it quickly became clear that roads in high and low latitudes met at odd angles on the map instead of the 90 degrees in reality.

As the primary purpose of Google Maps was street-level driving directions, Google sacrificed the global view accuracy for far better relative accuracy among streets when viewing a single city. Competing mapping systems followed suit. Google also standardized on the WGS 84 datum. This datum defines a specific spherical model of the Earth called a geoid. This model defines the normal sea level. What is significant about this choice by Google is that the **Global Positioning System** (**GPS**) also uses this datum. Therefore, most GPS units default to this datum as well, making Google Maps easily compatible with raw GIS data.

The Google variation of the Mercator projection is often called Google Mercator. The **European Petroleum Survey Group** (**EPSG**) assigns short numeric codes to projections as an easy way to reference them. Rather than waiting for the EPSG to approve or assign a code that was first only relevant to Google, they began calling the projection EPSG:900913, which is *Google* spelled with numbers. Later, EPSG assigned the code EPSG:3857, deprecating the older code. Most GIS systems recognize the two codes as synonymous. It should be noted that Google tweaked the standard Mercator projection slightly for its use; however, this variation is almost imperceptible. Google uses spherical formulas at all map scales while the standard Mercator assumes an ellipsoidal form at large scales.

The following URL provides an image taken from Wikipedia:

```
https://en.wikipedia.org/wiki/File:Tissot_mercator.png
```

It shows the distortion caused by the Mercator projection using Tissot's indicatrix, which projects small ellipses of equal size into a map. The distortion of the ellipse clearly shows how the projection affects the size and distance:

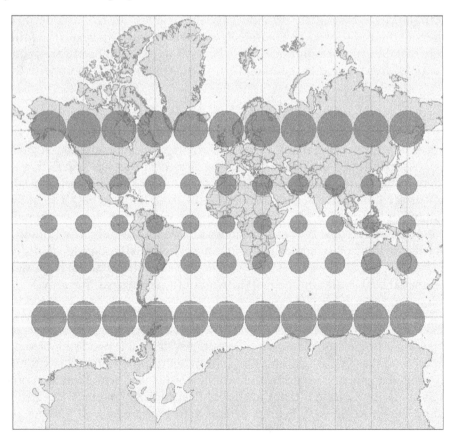

Web mapping services have reduced the chore of hunting for data and much of the preprocessing for analysts to create base maps. However, to create anything of value, you must understand geospatial data and how to work with it. This chapter provides an overview of the common data types and issues that you will encounter in geospatial analysis. Throughout this chapter, two terms will be commonly used: **vector data** and **raster data**. These are the two primary categories under which most geospatial datasets can be grouped. Vector data includes any format that minimally represents geolocation data using points, lines, or polygons. Raster data includes any format that stores data in a grid of rows and columns. Raster data includes all image formats.

If you want to see a projection that shows the relative size of continents more accurately, refer to the Goode homolosine projection:

```
https://en.wikipedia.org/wiki/Goode_homolosine_
projection
```

Data structures

Despite dozens of formats, geospatial data have common traits. Understanding these traits can help you approach and understand unfamiliar data formats by identifying the ingredients common to nearly all spatial data. The structure of a given data format is usually driven by its intended use. Some data is optimized for efficient storage or compression, some is optimized for efficient access, some is designed to be lightweight and readable (web formats), while other data formats seek to contain as many different data types as possible.

Interestingly, some of the most popular formats today are also some of the simplest and even lack features found in more capable and sophisticated formats. Ease of use is extremely important to geospatial analysts because so much time is spent integrating data into geographic information systems as well as exchanging data among analysts. Simple data formats facilitate these activities the best.

Common traits

Geospatial analysis is an approach applying information processing techniques to data with geographic context. This definition contains the most important elements of geospatial data: geolocation data and subject information. These two factors are present in every format that can be considered geospatial data. Another common feature of geospatial data is spatial indexing. Overview datasets are also related to indexing.

Geolocation

Geolocation information can be as simple as a single point on the Earth referencing where a photo was taken. It can also be as complex as a satellite camera engineering model and orbital mechanics information to reconstruct the exact conditions and location under which the satellite captured the image.

Subject information

Subject information can also cover a wide range of possibilities. Sometimes, the pixels in an image are the data in terms of a visual representation of the ground. At other times, an image may be processed using multispectral bands, such as infrared light, to provide information not visible in the image. Processed images are often classified using a structured color palette that is linked to a key, describing the information each color represents. Other possibilities include some form of database with rows and columns of information for each geolocated feature, such as the population associated with each city in our SimpleGIS from *Chapter 1, Learning Geospatial Analysis with Python*.

Spatial indexing

Geospatial datasets are often very large files easily reaching hundreds of megabytes or even several gigabytes in size. Geospatial software can be quite slow in trying to repeatedly access large files when performing analysis. As discussed briefly in *Chapter 1, Learning Geospatial Analysis with Python*, spatial indexing creates a guide, which allows software to quickly locate query results without examining every single feature in the dataset. Spatial indexes allow software to eliminate possibilities and perform more detailed searches or comparisons on a much smaller subset of the data.

Indexing algorithms

Many spatial indexing algorithms are derivatives of well-established algorithms used for decades on nonspatial information. The two most common spatial indexing algorithms are **Quadtree index** and **R-tree index**.

Quadtree index

The Quadtree algorithm actually represents a series of different algorithms based on a common theme. Each node in a Quadtree index contains four children. These child nodes are typically square or rectangular in shape. When a node contains a specified number of features, the node splits if additional features are added. The concept of dividing a space into nested squares speeds up spatial searches. Software must only handle five points at a time and use simple greater-than/less-than comparisons to check whether a point is inside a node. Quadtree indexes are most commonly found as file-based index formats.

The following image shows a point dataset sorted by a Quadtree algorithm. The black points are the actual dataset, while the boxes are the bounding boxes of the index. Note that none of the bounding boxes overlap. The left image shows the spatial representation of the index. The right image shows the hierarchical relationship of a typical index which is how spatial software sees the index and data. This structure allows a spatial search algorithm to quickly eliminate possibilities when trying to locate one or more points in relation to some other set of features.

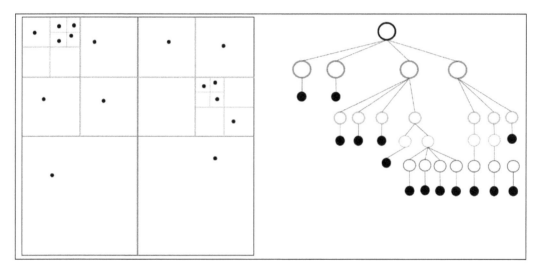

R-tree index

R-tree indexes are more sophisticated than Quadtrees. R-trees are designed to handle three-dimensional data and are optimized to store the index in a way that is compatible with the way databases use disk space and memory. Nearby objects are grouped together using any algorithm from a variety of spatial algorithms. All objects in a group are bounded by a minimum rectangle. These rectangles are aggregated into hierarchical nodes that are balanced at each level. Unlike a Quadtree, the bounding boxes of an R-tree may overlap across nodes. Due to the relative complexity and database-oriented structure, R-trees are most commonly found in spatial databases as opposed to file-based formats.

The following diagram from `https://en.wikipedia.org/wiki/File:R-tree.svg` shows a balanced R-tree for a two-dimensional point dataset:

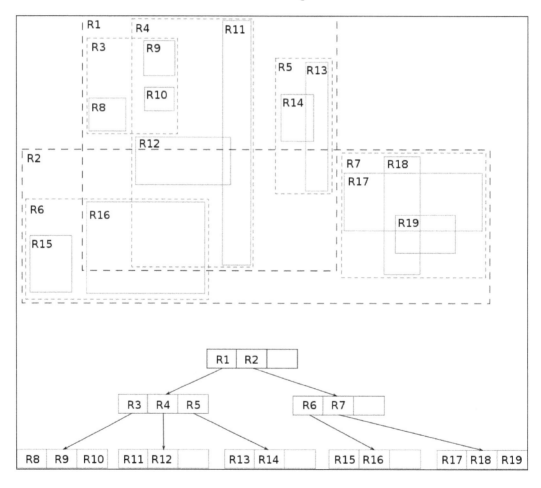

Grids

Spatial indexes also often employ the concept of an integer grid. Geospatial coordinates are usually floating point decimal numbers with anywhere from 2 to 16 decimal places. Performing comparisons on floating point numbers is far more computationally expensive than working with integers. Indexed searching is about first eliminating possibilities that don't require precision. Most spatial indexing algorithms, therefore, map floating point coordinates to a fixed-sized integer grid. On searching for a particular feature, the software can use more efficient integer comparisons rather than working with floating point numbers. Once the results are narrowed down, the software can access the full resolution data.

Grid sizes can be as small as 256 by 256 for simple file formats or can be as large as three million by three million in large geospatial databases designed to incorporate every known coordinate system and possible resolution. The integer mapping technique is very similar to the rendering technique that is used to plot data on a graphics canvas in mapping programs. The SimpleGIS script in *Chapter 1, Learning Geospatial Analysis with Python*, also uses this technique to render points and polygons using the built-in Python turtle graphics engine.

Overviews

Overview data is most commonly found in raster formats. Overviews are resampled, lower resolution versions of raster datasets that provide thumbnail views or simply faster loading image views at different map scales. They are also known as pyramids and the process of creating them is known as pyramiding an image. These overviews are usually preprocessed and stored with the full resolution data either embedded with the file or in a separate file. The compromise of this convenience is that the additional images add to the overall file size of the dataset; however, they speed up image viewers. Vector data also has a concept of overviews, usually to give a dataset geographic context in an overview map. However, because vector data is scalable, reduced size overviews are usually created on the fly by software using a generalization operation as mentioned in *Chapter 1, Learning Geospatial Analysis with Python*.

Occasionally, vector data is rasterized by converting it into a thumbnail image, which is stored with or embedded in the image header. The following image demonstrates the concept of image overviews that also shows visually why they are often called pyramids:

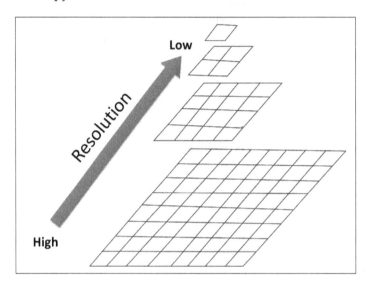

Metadata

As discussed in *Chapter 1, Learning Geospatial Analysis with Python*, metadata is any data that describes the associated dataset. Common examples of metadata include basic elements such as the footprint of the dataset on the Earth as well as more detailed information such as spatial projection and information describing how the dataset was created. Most data formats contain the footprint or bounding box of the data on the Earth. Detailed metadata is typically stored in a separate location in a standard format such as the U.S. **Federal Geographic Data Committee (FGDC) Content Standard for Digital Geospatial Metadata (CSDGM)**, ISO, or the newer European Union initiative, which includes metadata requirements, called the **Infrastructure for Spatial Information in the European Community (INSPIRE)**.

File structure

The preceding elements can be stored in a variety of ways in a single file, multiple files, or database depending on the format. Additionally, this geospatial information can be stored in a variety of formats, including embedded binary headers, XML, database tables, spreadsheets/CSV, separate text, or binary files.

Human readable formats such as XML files, spreadsheets, and structured text files require only a text editor to investigate. These files are also easily parsed and processed using Python's built-in modules, data types, and string manipulation functions. Binary-based formats are more complicated. It is thus typically easier to use a third-party library to deal with binary formats.

However, you don't have to use a third-party library, especially if you just want to investigate the data at a high level. Python's built-in struct module has everything that you need. The struct module lets you read and write binary data as strings. When using the struct module, you need to be aware of the concept of byte order. Byte order refers to how the bytes of information that make up a file are stored in memory. This order is usually platform-specific, but in some rare cases, including shapefiles, the byte order is mixed in the file. The Python struct module uses the greater than (>) and less than (<) symbols to specify byte order (big endian and little endian, respectively).

The following brief example demonstrates the usage of the Python struct module to parse the bounding box coordinates from an Esri shapefile vector dataset. You can download this shapefile as a zipped file at the following URL:

```
https://geospatialpython.googlecode.com/files/hancock.zip
```

When you unzip this, you will see three files. For this example, we'll be using `hancock.shp`. The Esri shapefile format has a fixed location and data type in the file header from byte 36 to byte 37 for the minimum x, minimum y, maximum x, and maximum y bounding box values. In this example, we will execute the following steps:

1. Import the `struct` module.
2. Open the `hancock.zip` shapefile in the binary read mode.
3. Navigate to byte `36`.
4. Read each 8-byte double specified as `d`, and unpack it using the `struct` module in little-endian order as designated by the < sign.

The best way to execute this script is in the interactive Python interpreter. We will read the minimum longitude, minimum latitude, maximum longitude, and maximum latitude:

```
>>> import struct
>>> f = open("hancock.shp","rb")
>>> f.seek(36)
>>> struct.unpack("<d", f.read(8))
(-89.6904544701547,)
>>> struct.unpack("<d", f.read(8))
(30.173943486533133,)
```

```
>>> struct.unpack("<d", f.read(8))
(-89.32227546981174,)
>>> struct.unpack("<d", f.read(8))
(30.6483914869749,)
```

You'll notice that when the `struct` module unpacks a value, it returns a Python tuple with one value. You can shorten the preceding unpacking code to one line by specifying all four doubles at once and increasing the byte length to 32 bytes as shown in the following code:

```
>>> f.seek(36)
>>> struct.unpack("<dddd", f.read(32))
(-89.6904544701547, 30.173943486533133, -89.32227546981174,
  30.6483914869749)
```

Vector data

Vector data is, by far, the most common geospatial format because it is the most efficient way to store spatial information, and in general, requires less computer resources to store and process than raster data. The OGC has over 16 formats directly related to vector data. Vector data stores only geometric primitives including points, lines, and polygons. However, only the points are stored for each type of shape. For example, in the case of a simple straight vector line shape, only the end points would be necessarily stored and defined as a line. Software displaying this data would read the shape type and then connect the end points with a line dynamically.

Geospatial vector data is similar to the concept of vector computer graphics with some notable exceptions. Geospatial vector data contains positive and negative Earth-based coordinates, while vector graphics typically store computer screen coordinates. Geospatial vector data is also usually linked to other information about the object represented by the geometry. This information may be as simple as a timestamp in the case of GPS data or an entire database table for larger geographic information systems. Vector graphics often store styling information describing colors, shadows, and other display-related instructions, while geospatial vector data typically does not. Another important difference is shapes. Geospatial vectors typically only include very primitive geometries based on points, straight lines, and straight-line polygons, while many computer graphic vector formats have concepts of curves and circles. However, geospatial vectors can model these shapes using more points.

Other human readable formats such as **Comma-Separated Values** (**CSV**), simple text strings, GeoJSON, and XML-based formats are technically vector data because they store geometry as opposed to rasters, which represent all the data within the bounding box of the dataset. Until the explosion of XML in the late 1990s, vector data formats were nearly all binary. XML provided a hybrid approach that was both computer and human readable. The compromise is that text formats such as GeoJSON and XML data greatly increase the file size over binary formats. These formats are discussed later in this section.

The number of vector formats to choose from is staggering. The open source vector library, OGR (`http://www.gdal.org/ogr_formats.html`), lists over 86 supported vector formats. Its commercial counterpart, Safe Software's **Feature Manipulation Engine** (**FME**), lists over 188 supported vector formats (`http://www.safe.com/fme/format-search/#filters%5B%5D=VECTOR`). These lists include a few vector graphics formats as well as human readable geospatial formats. There are still dozens of formats out there to at least be aware of, in case you come across them.

Shapefiles

The most ubiquitous geospatial format is the Esri shapefile. Geospatial software company Esri released the shapefile format specification as an open format in 1998 (`http://www.esri.com/library/whitepapers/pdfs/shapefile.pdf`). Esri developed it as a format for their ArcView software, designed as a lower-end GIS option to complement their high-end professional package, ArcInfo, formerly called ARC/INFO. However, the open specification, efficiency, and simplicity of the format turned it into an unofficial GIS standard, still extremely popular over 15 years later. Virtually, every piece of software labeled as geospatial software supports shapefiles because the shapefile format is so common. For this reason, you can almost get by as an analyst by being intimately familiar with shapefiles and mostly ignoring other formats. You can convert almost any other format to shapefiles through the source format's native software or a third-party converter, such as the OGR library, for which there is a Python module. Other Python modules to handle shapefiles are Shapely and Fiona, which are based on OGR.

One of the most striking features of a shapefile is that the format consists of multiple files. At a minimum, there are three, and there can even be as many as 15 different files! The following table describes the file formats. The `.shp`, `.shx`, and `.dbf` files are required for a valid shapefile.

Shapefile supporting file extension	Supporting file purpose	Notes
`.shp`	It is the shapefile. It contains the geometry.	Required file. Some software needing only geometry will accept the `.shp` files without the `.shx` or `.dbf` file.
`.shx`	It is the shape index file. It is fixed-sized record index referencing geometry for faster access.	Required file. This file is meaningless without the `.shp` file.
`.dbf`	It is the database file. It contains the geometry attributes.	Required file. Some software will access this format without the `.shp` file present as the specification predates shapefiles. Based on the very old FoxPro and Dbase formats. An open specification exists called Xbase. The `.dbf` files can be opened by most types of spreadsheet software.
`.sbn`	It is the spatial bin file, the shapefile spatial index.	Contains bounding boxes of features mapped to a 256-by-256 integer grid. Frequently seen.
`.sbx`	A fixed-sized record index for the `.sbn` file.	A traditional ordered record index of a spatial index. Frequently seen.

Shapefile supporting file extension	Supporting file purpose	Notes
.prj	Map projection information stored in well-known text format.	Very common file and required for on-the-fly projection by the GIS software. This same format can also accompany raster data.
.fbn	A spatial index of read only features.	Very rarely seen.
.fbx	A fixed-sized record index of the .fbn spatial index.	Very rarely seen.
.ixs	A geocoding index.	Common in geocoding applications including driving-direction type applications.
.mxs	Another type of geocoding index.	Less common than the .ixs format.
.ain	Attribute index.	Mostly legacy format and rarely used in modern software.
.aih	Attribute index.	Accompanies the .ain files.
.qix	Quadtree index.	A spatial index format created by the open source community because the Esri .sbn and .sbx files were undocumented until recently.
.atx	Attribute index.	A more recent Esri software-specific attribute index to speed up attribute queries.

Shapefile supporting file extension	Supporting file purpose	Notes
`.shp.xml`	Metadata.	Geospatial metadata .xml container. Can be any of the multiple XML standards including FGDC and ISO.
`.cpg`	Code page file for `.dbf`.	Used for the internationalization of the `.dbf` files.

You will probably never encounter all of these formats at once. However, any shapefile that you use will have multiple files. You will commonly see `.shp`, `.shx`, `.dbf`, `.prj`, `.sbn`, `.sbx`, and, occasionally, `.shp.xml` files. If you want to rename a shapefile, you must rename all of the associated files with the same name; however, in Esri software and other GIS packages, these datasets will appear as a single file.

Another important feature of shapefiles is that the records are not numbered. Records include the geometry, the `.shx` index record, and the `.dbf` record. These records are stored in a fixed order. When you examine shapefile records using software, they appear to be numbered. People are often confused when they delete a shapefile record, save the file, and reopen it; the number of the record deleted still appears. The reason is that the shapefile records are numbered dynamically on loading but not saved. So, if you delete record number 23 and save the shapefile, record number 24 will become 23 the next time you read the shapefile. Many people expect to open the shapefile and see the records jump from 22 to 24. The only way to track shapefile records in this way is to create a new attribute called ID or similar in the `.dbf` file and assign each record a permanent, unique identifier.

Just like renaming shapefiles, care must be taken when editing shapefiles. It's best to use software that treats the shapefiles as a single dataset. If you edit any of the files individually and add or delete a record without editing the accompanying files, the shapefile will be seen as corrupt by most geospatial software.

CAD files

CAD stands for **computer-aided design**. The primary formats for CAD data were created by Autodesk for their leading AutoCAD package. The two formats commonly seen are **Drawing Exchange Format (DXF)** and AutoCAD native **Drawing (DWG)** format. DWG was traditionally a closed format but it has become more open.

CAD software is used for everything that is engineering-related, from designing bicycles to cars, parks, and city sewer systems. As a geospatial analyst, you don't have to worry about mechanical engineering designs; however, civil engineering designs become quite an issue. Most engineering firms use geospatial analysis to a very limited degree but store nearly all of their data in the CAD format. The DWG and DXF formats can represent objects using features not found in geospatial software or weakly supported by geospatial systems. Some examples of these features include the following:

- Curves
- Surfaces (for objects that are different from geospatial elevation surfaces)
- 3D solids
- Text (rendered as an object)
- Text styling
- Viewport configuration

These CAD and engineering-specific features make it difficult to cleanly convert CAD data into geospatial formats. If you encounter CAD data, the easiest option is to ask the data provider if they have shapefiles or some other geospatial-centric format.

Tag-based and markup-based formats

Tag-based markup formats are typically XML formats. They also include other structured text formats such as the **well-known text** (**WKT**) format used for projection information files as well as different types of data exchange. XML formats include **Keyhole Markup Language** (**KML**), the **Open Street Map** (**OSM**) format, and the Garmin GPX format for GPS data, which has become a popular exchange format. The Open Geospatial Consortium's **Geographic Markup Language** (**GML**) standard is one of the oldest and most widely used XML-based geographic formats. It is also the basis for the OGC **Web Feature Service** (**WFS**) standard for web applications. However, GML has been largely superseded by KML and the GeoJSON format.

XML formats often contain more than just geometry. They also contain attributes and rendering instructions such as color, styling, and symbology. Google's KML format has become a fully supported OGC standard. The following is a sample of KML showing a simple place mark:

```
<?xml version="1.0" encoding="utf-8"?>
<kml xmlns="http://www.opengis.net/kml/2.2">
  <Placemark>
    <name>Mockingbird Cafe</name>
    <description>Coffee Shop</description>
```

```
      <Point>
        <coordinates>-89.329160,30.310964</coordinates>
      </Point>
    </Placemark>
  </kml>
```

The XML format is attractive to geospatial analysts for the following several reasons:

- It is a human readable format
- It can be edited in a text editor
- It is well-supported by programming languages (especially Python!)
- It is, by definition, easily extensible

XML is not perfect though. It is an inefficient storage mechanism for very large data formats and can quickly become cumbersome to edit. Errors in datasets are common and most parsers do not handle errors robustly. Despite the downsides, XML is widely used in geospatial analysis.

Scalable Vector Graphics (SVG) is a widely supported XML format for computer graphics. It is supported well by browsers and is often used for geospatial rendering. However, SVG was not designed as a geographic format.

The WKT format is also an older OGC standard. The most common use for it is to define projection information usually stored in the .prj projection files along with a shapefile or raster. The WKT string for the WGS84 coordinate system is as follows:

```
GEOGCS["WGS 84",
    DATUM["WGS_1984",
        SPHEROID["WGS 84",6378137,298.257223563,
            AUTHORITY["EPSG","7030"]],
        AUTHORITY["EPSG","6326"]],
    PRIMEM["Greenwich",0,
        AUTHORITY["EPSG","8901"]],
    UNIT["degree",0.01745329251994328,
        AUTHORITY["EPSG","9122"]],
    AUTHORITY["EPSG","4326"]]
```

The parameters defining a projection can be quite long. A standards committee created by the EPSG created a numerical coding system to reference projections. These codes, such as EPSG:4326, are used as shorthand for strings like the preceding code. There are also short names for commonly used projections such as Mercator, which can be used in different software packages to reference a projection. More information on these reference systems can be found at the spatial reference website at http://spatialreference.org/ref/.

GeoJSON

GeoJSON is a relatively new and brilliant text format based on the **JavaScript Object Notation (JSON)** format, which has been a commonly used data exchange format for years. Despite its short history, GeoJSON can be found embedded in all major geospatial software systems and most websites that distribute data because JavaScript is the language of the dynamic web and GeoJSON can be directly fed into JavaScript.

GeoJSON is a completely backwards-compatible extension for the popular JSON format. The structure of JSON is very similar and in some cases identical to existing data structures of common programming languages. JSON is almost identical to Python's dictionary and list data types. Due to this similarity, parsing JSON in a script is simple to do from scratch but there are also many libraries to make it even easier. Python contains a built-in library aptly named `json`.

GeoJSON provides you with a standard way to define geometry, attributes, bounding boxes, and projection information. GeoJSON has all of the advantages of XML including human readable syntax, excellent software support, and wide use in the industry. It also surpasses XML. GeoJSON is far more compact than XML largely because it uses simple symbols to define objects rather than opening and closing text-laden tags. The compactness also helps with the readability and manageability of larger datasets. However, it is still inferior to binary formats from a data volume standpoint. The following is a sample of the GeoJSON syntax, defining a geometry collection with both a point and line:

```
{ "type": "GeometryCollection",
  "geometries": [
    { "type": "Point",
      "coordinates": [-89.33, 30.0]
    },
    { "type": "LineString",
      "coordinates": [ [-89.33, 30.30], [-89.36, 30.28] ]
    }
  {"type": "Polygon",
    "coordinates": [[
      [-104.05, 48.99],
      [-97.22,  48.98]
    }
  ]
}
```

The preceding code is a valid GeoJSON, but it is also a valid Python data structure. You can copy the preceding code sample directly to the Python interpreter as a variable definition and it will evaluate without error as follows:

```
gc = { "type": "GeometryCollection",
   "geometries": [
     { "type": "Point",
       "coordinates": [-89.33, 30.0]
     },
     { "type": "LineString",
       "coordinates": [ [-89.33, 30.30], [-89.36, 30.28] ]
     }
   ]
}
gc
{'type': 'GeometryCollection', 'geometries': [{'type': 'Point',
   'coordinates': [
   -89.33, 30.0]}, {'type': 'LineString', 'coordinates': [[-89.33,
   30.3], [-89.36,30.28]]}]}
```

Due to its compact size, Internet-friendly syntax by virtue of being written in JavaScript, and support from major programming languages, GeoJSON is a key component of leading REST geospatial web APIs, which will be covered later in this chapter. It currently offers the best compromise among the computer resource efficiency of binary formats, the human-readability of text formats, and programmatic utility.

Raster data

Raster data consists of rows and columns of cells or pixels, with each cell representing a single value. The easiest way to think of raster data is as images, which is how they are typically represented by software. However, raster datasets are not necessarily stored as images. They can also be ASCII text files or **Binary Large Objects (BLOBs)** in databases.

Another difference between geospatial raster data and regular digital images is resolution. Digital images express resolution as dots-per-inch if printed in full size. Resolution can also be expressed as the total number of pixels in the image defined as megapixels. However, geospatial raster data uses the ground distance that each cell represents. For example, a raster dataset with a two-foot resolution means that a single cell represents two feet on the ground, which also means that only objects larger than two feet can be identified visually in the dataset.

Raster datasets may contain multiple bands, meaning that different wavelengths of light can be collected at the same time over the same area. Often, this range is from 3-7 bands but can be several hundred in hyperspectral systems. These bands are viewed individually or swapped in and out as the RGB bands of an image. They can also be recombined into a derivative single-band image using mathematics and then recolored using a set number of classes representing like values within the dataset.

Another common application of raster data is in the field of scientific computing which shares many elements of geospatial remote sensing but adds some interesting twists. Scientific computing often uses complex raster formats, including **Network Common Data Form (NetCDF)**, **GRIB**, and **HDF5**, which store entire data models. These formats are more like directories in a filesystem and can contain multiple datasets or multiple versions of the same dataset. Oceanography and meteorology are the most common applications of this kind of analysis. An example of a scientific computing dataset is the output of a weather model, where the cells of the raster dataset in different bands may represent different variables' output from the model in a time series.

Like vector data, raster data can come in a variety of formats. The open source raster library called **Geospatial Data Abstraction Library (GDAL)**, which actually includes the vector OGR library mentioned earlier, lists over 130 supported raster formats (`http://www.gdal.org/formats_list.html`). The FME software package supports this many as well. However, just like shapefiles and CAD data, there are a few standout raster formats.

TIFF files

The **Tagged Image File Format (TIFF)**, is the most common geospatial raster format. The TIFF format's flexible tagging system allows it to store any type of data, whatsoever, in a single file. TIFFs can contain overview images, multiple bands, integer elevation data, basic metadata, internal compression, and a variety of other data typically stored in additional supporting files by other formats. Anyone can extend the TIFF format unofficially by adding tagged data to the file structure. This extensibility has benefits and drawbacks, however. A TIFF file may work fine in one piece of software but fails when accessed in another because the two software packages implement the massive TIFF specification to different degrees. An old joke about TIFFs has a frustrating amount of truth to it: TIFF stands for *Thousands of Incompatible File Formats*. The GeoTIFF extension defines how geospatial data is stored. Geospatial rasters stored as TIFF files may have any of the following file extensions: `.tiff`, `.tif`, or `.gtif`.

JPEG, GIF, BMP, and PNG

JPEG, GIF, BMP, and PNG formats are common image formats, in general, but can be used for basic geospatial data storage as well. Typically, these formats rely on accompanying the supporting text files for the georeferencing of the information in order to make them compatible with the GIS software such as WKT, .prj, or world files described here.

The JPEG format is also fairly common for geospatial data. JPEGs have a built-in metadata tagging system similar to TIFFs called EXIF. JPEGs are commonly used for geotagged photographs in addition to raster GIS layers. **Bitmap (BMP)** images are used for desktop applications and document graphics. However, JPEG, GIF, and PNG are the formats used in web mapping applications, especially for server pregenerated map tiles for quick access via slippy maps.

Compressed formats

As geospatial rasters tend to be very large, they are often stored using advanced compression techniques. The latest open standard is the JPEG 2000 format, which is an update of the JPEG format and includes wavelet compression and a few other features such as georeferencing data. **Multi-resolution Seamless Image Database (MrSID)** (.sid) and **Enhanced Compression Wavelet (ECW)** (.ecw) are two proprietary wavelet compression formats often seen in geospatial contexts. The TIFF format supports compression including the **Lempel-Ziv-Welch (LZW)** algorithm. It must be noted that compressed data is suitable as part of a base map but should not be used for remote sensing processing. Compressed images are designed to look visually correct but often alter the original cell value. Lossless compression algorithms try to avoid degrading the source data but it's generally considered a bad idea to attempt spectral analysis on data that has been through compression. The JPEG format is designed to be a lossy format that sacrifices data for a smaller file size. It is also commonly encountered so it is important to remember this fact to avoid invalid results.

ASCII Grids

Another means of storing raster data, often elevation data, is in ASCII Grid files. This file format was created by Esri but has become an unofficial standard supported by most software packages. An ASCII Grid is a simple text file containing (x, y) values as rows and columns. The spatial information for the raster is contained in a simple header. The format of the file is as follows:

```
<NCOLS xxx>
<NROWS xxx>
```

```
<XLLCENTER xxx  |  XLLCORNER xxx>
<YLLCENTER xxx  |  YLLCORNER xxx>
<CELLSIZE xxx>
{NODATA_VALUE xxx}
row 1
row 2
.
.
.
row n
```

While not the most efficient way to store data, ASCII Grid files are very popular because they don't require any special data libraries to create or access geospatial raster data. These files are often distributed as zip files. The header values in the preceding format contain the following information:

- The number of columns
- The number of rows
- The x-axis cell center coordinate | x-axis lower-left corner coordinate
- The y-axis cell center coordinate | y-axis lower-left corner coordinate
- The cell size in mapping units
- The no-data value (typically, 9999)

World files

World files are simple text files, which can provide geospatial referencing information to any image externally for file formats that typically have no native support for spatial information, including JPEG, GIF, PNG, and BMP. The world file is recognized by geospatial software due to its naming convention. The most common way to name a world file is to use the raster file name and then alter the extension to remove the middle letter and add w at the end. The following table shows some examples of raster images in different formats and the associated world file name based on the convention:

Raster file name	World file name
World.jpg	World.jpw
World.tif	World.tfw
World.bmp	World.bpw
World.png	World.pgw
World.gif	World.gfw

The structure of a world file is very simple. It is a six-line text file as follows:

- Line 1: The cell size along the x axis in ground units
- Line 2: The rotation on the y axis
- Line 3: The rotation on the x axis
- Line 4: The cell size along the y axis in ground units
- Line 5: The center x-coordinate of the upper left cell
- Line 6: The center y-coordinate of the upper left cell

The following is an example of world file values:

```
15.0
0.0
0.0
-15.0
-89,38
45.0
```

The (x, y) coordinates and the (x, y) cell size contained in lines 1, 4, 5, and 6 allow you to calculate the coordinate of any cell or the distance across a set of cells. The rotation values are important for geospatial software because remotely sensed images are often rotated due to the data collection platform. Rotating the images runs the risk of resampling the data and, therefore, data loss so the rotation values allow the software to account for the distortion. The surrounding pixels outside the image are typically assigned a no data value and represented as the color black. The following image, courtesy of the **U.S. Geological Survey (USGS)**, demonstrates image rotation, where the satellite collection path is oriented from southeast to northeast but the underlying base map is north:

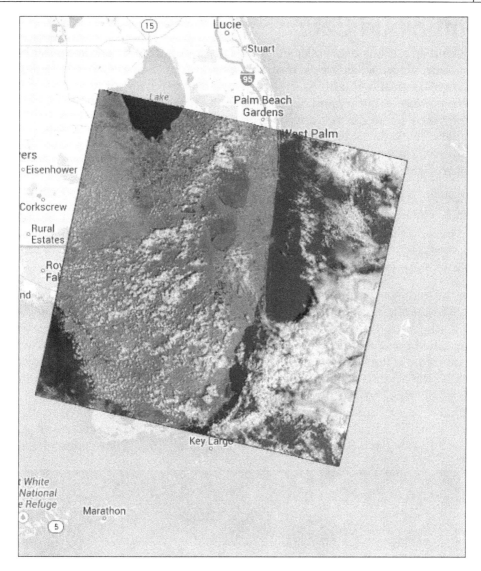

World files are a great tool when working with raster data in Python. Most geospatial software and data libraries support world files so they are usually a good choice for the georeferencing.

You'll find that world files are very useful, but as you use them infrequently, you forget what the unlabeled contents represent. A quick reference for world files is available at http://kralidis.ca/gis/worldfile.htm.

Point cloud data

Point cloud data is any data collected as the (x, y, z) location of a surface point based on some sort of focused energy return. Point cloud data can be created using lasers, radar waves, acoustic soundings, or other waveform generation devices. The spacing between points is arbitrary and dependent on the type and position of the sensor collecting the data. In this book, we will primarily be concerned with LIDAR data and radar data. Radar point cloud data is typically collected on space missions while LIDAR is typically collected by terrestrial or airborne vehicles. Conceptually, both the types of data are similar.

LIDAR uses powerful laser range-finding systems to model the world with very high precision. The term LIDAR or LiDAR is a combination of the words light and radar. Some people claim it also stands for **Light Detection and Ranging**. LIDAR sensors can be mounted on aerial platforms including satellites, airplanes, or helicopters. They can also be mounted on vehicles for ground-based collection.

Due to the high-speed, continuous data collection provided by LIDAR, and a wide field of view—often 360 degrees of the sensor—LIDAR data doesn't typically have a rectangular footprint the way other forms of raster data do. LIDAR datasets are usually called point clouds because the data is a stream of (x,y,z) locations, where z is the distance from the laser to a detected object and the (x,y) values are the projected location of the object calculated from the location of the sensor. The following image, courtesy of USGS, shows a point cloud LIDAR dataset in an urban area using a terrestrial sensor as opposed to an aerial one. The colors are based on the strength of the laser's energy return with red areas being closer to the LIDAR sensor and green areas farther away, which can give a precise height to within a few centimeters:

The most common data format for LIDAR data is **LIDAR Exchange Format (LAS)**, which is a community standard. LIDAR data can be represented in many ways including a simple text file with one (x, y, z) tuple per line. Sometimes, LIDAR data can be colorized using image pixel colors collected at the same time. LIDAR data can also be used to create 2D elevation rasters. This technique is the most common use for LIDAR in geospatial analysis. Any other use requires specialized software that allows the user to work in 3D. In that case, other geospatial data cannot be combined with the point cloud.

Web services

Geospatial web services allow users to perform data discovery, data visualization, and data access across the web. Web services are usually accessed by applications based on user input such as zooming an online map or searching a data catalogue. The most common protocols are the **Web Map Service (WMS)** that returns a rendered map image and **Web Feature Service (WFS)** that typically returns GML, which was mentioned in this chapter's introduction. Many WFS services can also return KML, JSON, zipped shapefiles, and other formats. These services are called through HTTP GET requests. The following URL is an example of a WMS GET request, which returns a map image of the world that is 640 pixels wide and 400 pixels tall and an EPSG code of 900913:

```
http://osm.woc.noaa.gov/mapcache?SERVICE=wms&VERSION=1.1.1&REQUEST=Ge
tMap&FORMAT=image/png&WIDTH=600&HEIGHT=400&LAYERS=osm&SRS=EPSG:900913
&BBOX=-20037508,-20037508,20037508,20037508
```

Web services are rapidly evolving. The Open GIS Consortium is adding new standards for sensor networks and other geospatial contexts. **Representational State Transfer (REST)** services are also commonly used. REST services use simple URLs to make the requesting of data very easy to implement in nearly any programming language by tailoring URL parameters and their values accordingly. Nearly every programming language has robust HTTP client libraries capable of using the REST services. The REST services can return many types of data including images, XML, or JSON. There is no overarching geospatial REST standard yet, but the OGC has been working on one for quite some time. Esri has created a working implementation that is currently widely used. The following URL is an example of an Esri geospatial REST service that would return KML based on a weather radar image layer. You can add this URL to Google Earth as a network link or you can download it as compressed KML (KMZ) in a browser to import to another program:

```
http://gis.srh.noaa.gov/arcgis/rest/services/RIDGERadar/MapServer/gen
erateKml?docName=NWSRadar&layers=0&layerOptions=separateImage
```

You can find tutorials on the myriad of OGC services here: http://www.ogcnetwork.net/tutorials

Summary

You now have the background needed to work with common types of geospatial data. You also know the common traits of geospatial datasets that will allow you to evaluate unfamiliar types of data and identify key elements that will drive you towards which tools to use when interacting with this data. In *Chapter 3, The Geospatial Technology Landscape*, we'll examine the modules and libraries available to work with these datasets. As with all the code in this book, whenever possible, we'll use only pure Python and Python standard libraries.

3
The Geospatial Technology Landscape

The geospatial technology ecosystem consists of hundreds of software libraries and packages. This vast array of choices is overwhelming for newcomers to geospatial analysis. The secret to learning geospatial analysis quickly is to understand the handful of libraries and packages that really matter. Most software, both commercial and open source, is derived from these critical packages. Understanding the ecosystem of geospatial software and how it's used allows you to quickly comprehend and evaluate any geospatial tool.

Geospatial libraries can be assigned to one or more of the following high-level core capabilities, which they implement to some degree:

- Data access
- Computational geometry (including data reprojection)
- Visualization
- Metadata tools

Another important category is image processing for remote sensing; however, this category is very fragmented, containing dozens of software packages which are rarely integrated into derivative software. Most image processing software for remote sensing is based on the same data access libraries with custom image processing algorithms implemented on top of them. Take a look at the following examples of this type of software, which include both open source and commercial packages:

- **Open Source Software Image Map (OSSIM)**
- **Geographic Resources Analysis Support System (GRASS)**

- **Orfeo ToolBox (OTB)**
- **ERDAS IMAGINE**
- **ENVI**

Data access libraries such as GDAL and OGR are mostly written in either C or C++ for speed and cross-platform compatibility. Speed is important due to the commonly large sizes of geospatial datasets. However, you will also see many packages written in Java. When it's well-written, pure Java can approach speeds acceptable for processing large vector or raster datasets and are usually acceptable for most applications.

The following concept map shows the major geospatial software libraries and packages and how they are related. The libraries in bold represent root libraries that are actively maintained and not significantly derived from any other libraries. These root libraries represent geospatial operations, which are sufficiently difficult to implement, and the vast majority of people choose to use one of these libraries rather than create a competing one. As you can see, a handful of libraries make up a disproportionate amount of geospatial analysis software. And the following diagram is by no means exhaustive. In this book, we'll discuss only the most commonly used packages:

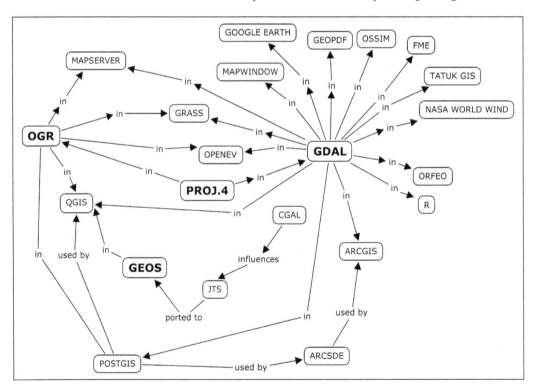

The libraries GDAL, OGR, GEOS, and PROJ.4 are the heart and soul of the geospatial analysis community on both the commercial and open source side. It is important to note that these libraries are all written in C or C++. There is also significant work done in Java in the form of the GeoTools and JTS core libraries, which are used across a range of desktops, servers, and mobile software. Given there are hundreds of geospatial packages available and nearly all relying on these libraries to do anything meaningful, you begin to get an idea of the complexity of geospatial data access and computational geometry. Compare this software domain to that of text editors, which return over 5,000 options when searched on the open source project site http://sourceforge.net/.

Geospatial analysis is a truly worldwide community with significant contributions to the field coming from every corner of the globe. But as you learn more about the heavy-hitting packages at the center of the software landscape, you'll see that these programs tend to come from Canada or are contributed heavily by Canadian developers. Credited as the birthplace of modern GIS, geospatial analysis is a matter of national pride. Also, the Canadian government and the public-private GeoConnections program have invested heavily in research and companies both to fuel the industry for economic reasons and out of necessity, to better manage the country's vast natural resources and the needs of its population.

In this chapter, we examine the packages which have had the largest impact on geospatial analysis and also those that you are likely to frequently encounter. However, as with any filtering of information, you are encouraged to do your own research and draw your own conclusions. The following websites offer more information on software not included in this chapter:

- Wikipedia list of GIS software: https://en.wikipedia.org/wiki/List_of_geographic_information_systems_software

- OSGeo project list and incubator projects: http://www.osgeo.org/

- FreeGIS.org software database: http://freegis.org/database/?cat=0&_ZopeId=18853465A58XbR1fIKo

Data access

As described in *Chapter 2*, *Geospatial Data*, geospatial datasets are typically large, complex, and varied. This challenge makes libraries, which efficiently read and in some cases, write this data essential to geospatial analysis. These libraries are also the most important. Without access to data, geospatial analysis doesn't begin. Furthermore, accuracy and precision are key factors in geospatial analysis. An image library that resamples data without permission, or a computational geometry library that rounds a coordinate even a couple of decimal places, can adversely affect the quality of analysis. Also, these libraries must manage memory efficiently. A complex geospatial process can last for hours or even days. If a data access library has a memory fault, it can delay an entire project or even an entire workflow involving dozens of people dependent on the output of that analysis.

GDAL

The **Geospatial Data Abstraction Library (GDAL)** does the most heavy lifting task in the geospatial industry. The GDAL website lists over 80 pieces of software using the library, and this list is by no means complete. Many of these packages are industry leading, open source, and commercial tools. This list doesn't include hundreds of smaller projects and individual analysts using the library for geospatial analysis. GDAL also includes a set of command-line tools that can do a variety of operations without any programming.

 A list of projects using GDAL can be found at the following URL:
`http://trac.osgeo.org/gdal/wiki/SoftwareUsingGdal`

GDAL provides a single, abstract data model for the vast array of raster data types found in the geospatial industry. It consolidates unique data access libraries for different formats and provides a common API for reading and writing data. Before developer Frank Warmerdam created GDAL in the late 1990s, each data format required a separate data access library with a different API to read data or the worse situation was that developers often wrote custom data access routines.

The following diagram provides a visual description of how GDAL abstracts raster data:

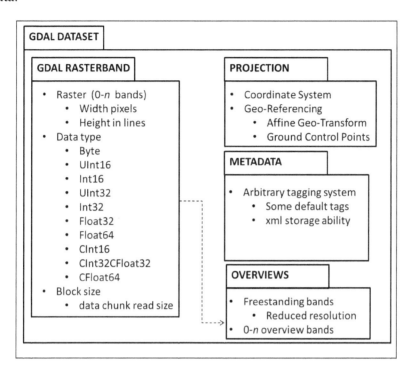

In the software concept map earlier in this chapter, you can see that GDAL has had the greatest impact of any single piece of geospatial software. Combine GDAL with its sister library OGR for vector data and the impact almost doubles. The PROJ.4 library has also had tremendous impact, but it is usually accessed via OGR or GDAL.

The GDAL homepage can be found at http://www.gdal.org/.

OGR

The OGR Simple Features Library is the vector data companion of GDAL. The OGR lists at least partial support for over 70 vector data formats. OGR originally stood for *OpenGIS Simple Features Reference Implementation*; however, it did not evolve into a reference implementation for the *Simple Features* standard even though the name stuck.

OGR serves the same purpose for vector data as GDAL does for raster data. It is also almost equally prolific in the geospatial industry. A part of the success of the GDAL/OGR package is the X11/MIT open source license. This license is both commercial and open source friendly. The GDAL/OGR library can be included in the proprietary software without revealing proprietary source code to users.

OGR has the following capabilities:

- Uniform vector data and modeling abstraction
- Vector data re-projection
- Vector data format conversion
- Attribute data filtering
- Basic geometry filtering including clipping and point-in-polygon testing

Like GDAL, OGR has several command-line utility programs, which demonstrate its capability. This capability can also be accessed through its programming API. The following diagram outlines the OGR architecture:

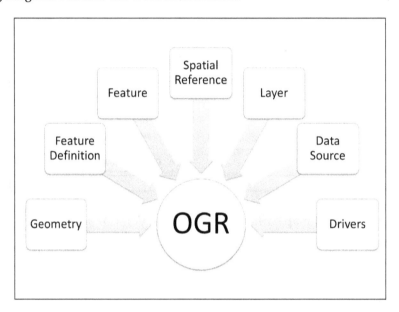

The OGR architecture is fairly concise, considering this model is able to represent over 70 different data formats. The Geometry object represents the OGC Simple Features Specification data model for points, linestrings, polygons, geometrycollections, multipolygons, multipoints, and multilinestrings. The Feature Definition object contains the attribute definitions of a group of related features. The Feature object ties the Geometry and Feature Definition information together. The Spatial Reference object contains an OGC Spatial Reference definition. The Layer object represents features grouped as layers within a data source. The Data Source is the file or database object accessed by OGR. The Driver object contains the translators for the 70 plus data formats available to OGR.

This architecture works smoothly with one minor quirk. The Layer concept is used even for data formats that only contain a single layer. For example, shapefiles can only represent a single layer. But, when you access a shapefile using OGR, on opening the data source, you must still invoke a new Layer object using the base name of the shapefile without a file extension. The design feature is only a minor inconvenience heavily outweighed by the power that OGR provides.

Computational geometry

Computational geometry encompasses the algorithms needed to perform operations on vector data. The field is very old in computer science; however, most of the libraries used for geospatial operations are separate from computer graphics libraries because of geospatial coordinate systems. As described near the end of *Chapter 1, Learning Geospatial Analysis with Python*, computer screen coordinates are almost always expressed in positive numbers, while geospatial coordinate systems often use negative numbers when they're moving west and south.

Several different geospatial libraries fit into the category but serve a wide range of uses from spatial selection to rendering. It should be noted that some features of the OGR Simple Features Library described previously move it beyond the category of data access and into the realm of computational geometry. But, it was included in the prior category because that is its primary purpose.

Computational geometry is a fascinating subject. While writing a simple script to automate a geospatial operation, you inevitably need some spatial algorithm. The question then arises that "Do you try to implement this algorithm yourself or go through the overhead of using a third-party library?" The choice is always deceptive because some tasks are visually easy to understand and implement; some look complex but turn out to be easy, and some are trivial to comprehend but are extraordinarily difficult. One such example is a geospatial buffer operation. The concept is easy enough, but the algorithm turns out to be quite difficult. The following libraries in this section are the major packages for computational geometry algorithms.

The PROJ.4 projection library

U.S. Geological Survey (USGS) analyst Jerry Evenden created what is now the PROJ.4 projection library in the mid 1990s while working at the USGS. Since then, it has become a project of the Open Source Geospatial Foundation with contributions from many other developers. PROJ.4 accomplishes the herculean task of transforming data among thousands of coordinate systems. The math to convert points among so many coordinate systems is extremely complex. No other library comes close to the capability PROJ.4. That fact and the routine needed by applications to convert data sets from different sources to a common projection make PROJ.4 the undisputed leader in this area.

The following plot is an example of how specific the projections supported by PROJ.4 can be. This image from OSGeo.org represents the Line/Station coordinate system of the **California Cooperative Oceanic Fisheries Investigations (CalCOFI)** program pseudo-projection used only by NOAA, the University of California Scripps Institution of Oceanography and the California Department of Fish and Wildlife to collect oceanographic and fisheries data over the last 60 years along the California coastline:

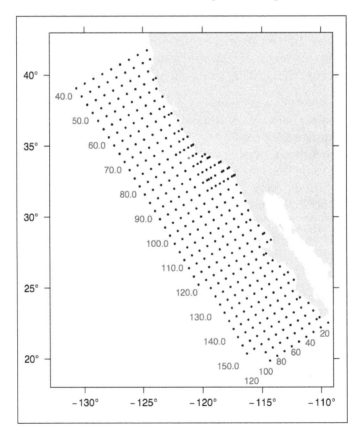

PROJ.4 uses a simple syntax capable of describing any projection including custom, localized ones as shown in the previous diagram. PROJ.4 can be found in virtually every major GIS package, which provides reprojection support and has its own command-line tools as well. It is available through both GDAL and OGR for vector and raster data. However, it is often useful to access the library directly because it gives you the ability to reproject individual points. Most of the libraries that incorporate PROJ.4 only let you reproject entire data sets.

For more information on PROJ.4 visit `https://github.com/OSGeo/proj.4/wiki`.

CGAL

The **Computational Geometry Algorithms Library** (**CGAL**), originally released in the late 1990s, is a robust and well-established open source computational geometry library. It is not specifically designed for geospatial analysis but is commonly used in the field.

CGAL is often referenced as a source for reliable geometry processing algorithms. The following image from the *CGAL User and Reference Manual* provides a visualization of one of the often referenced algorithms from CGAL called a polygon straight skeleton needed to accurately grow or shrink a polygon:

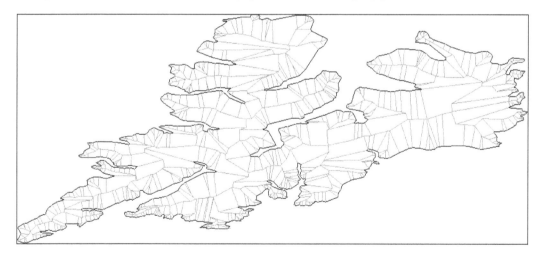

The straight skeleton algorithm is complex and important because shrinking or growing a polygon isn't just a matter of making it bigger or smaller. The polygon actually changes shape. As a polygon shrinks, non-adjacent edges collide and eliminate connecting edges. As a polygon grows, adjacent edges separate and new edges are formed to connect them. This process is key to buffering geospatial polygons. The following image, also from the *CGAL User and Reference Manual*, shows this effect using insets on the preceding polygon:

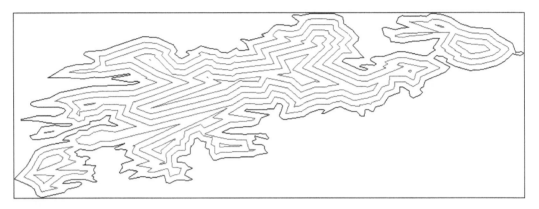

CGAL can be found online at http://www.cgal.org/.

JTS

The **Java Topology Suite (JTS)** is a geospatial computational geometry library written in 100 percent pure Java. JTS separates itself from other computational geometry libraries by implementing the **Open Geospatial Consortium (OGC)** Simple Features Specification for SQL. Interestingly, other developers have ported JTS to other languages including C++, Microsoft .NET, and even JavaScript.

JTS includes a fantastic test program called the JTS Test Builder, which provides a GUI to test functions without setting up an entire program. One of the most frustrating aspects of geospatial analysis concerns bizarre geometry shapes that break algorithms, which work most of the time. Another common issue is unexpected results due to tiny errors in data such as polygons that intersect themselves in very small areas that are not easily visible. The JTS Test Builder lets you interactively test JTS algorithms to verify data or just visually understand a process, as shown here:

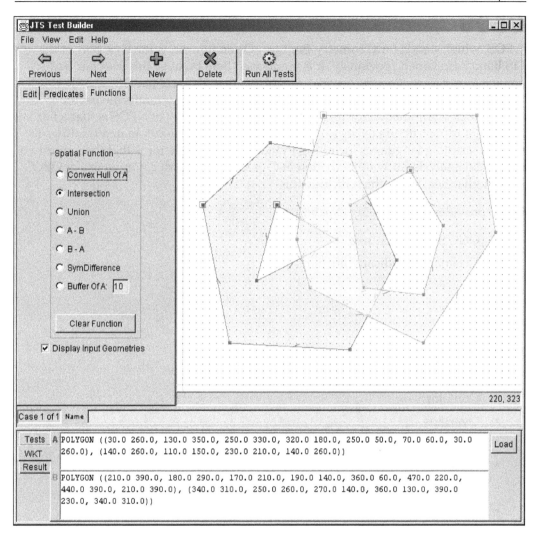

This tool is handy even if you aren't using JTS but one of the several ports to another language. It should be noted that Vivid Solutions, the maintainer of JTS, hasn't released a new version since JTS Version 1.8 in December 2006. The package is quite stable and still in active use. The JTS homepage is available at `http://www.vividsolutions.com/jts/JTSHome.htm`.

GEOS

GEOS, which stands for **Geometry Engine – Open Source**, is the C++ port of the JTS library explained previously. It is mentioned here because this port has had a much larger impact on the geospatial analysis than the original JTS. The C++ version can be compiled on many platforms as it avoids any platform-specific dependencies. Another factor responsible for the popularity of GEOS is that a fair amount of infrastructure exists to create automated or semi-automated bindings to various scripting languages including Python. One more factor is that the majority of geospatial analysis software is written in C or C++. The most common use of GEOS is through other APIs, which include this.

GEOS provides the following capabilities:

- OGC Simple Features
- Geospatial predicate functions
- Intersects
- Touches
- Disjoint
- Crosses
- Within
- Contains
- Overlaps
- Equals
- Covers
- Geospatial operations
- Union
- Distance
- Intersection
- Symmetric difference
- Convex hull
- Envelope
- Buffer
- Simplify

- Polygon assembly

- Polygon validation

- Area

- Length

- Spatial indexing

- OGC **well-known text (WKT)** and **well-known binary (WKB)** input/output

- C and C++ API

- Thread safety

GEOS can be compiled with GDAL to give OGR all of its capability. GEOS can be found online at `https://trac.osgeo.org/geos/`.

PostGIS

As far as open source geospatial databases go, PostGIS is the most commonly used spatial database. PostGIS is essentially a module on top of the well-known PostgreSQL relational database. Much of the power of PostGIS comes from the GEOS library mentioned earlier. Like JTS, it also implements the OGC Simple Features Specification for SQL. The combination of computational geometry ability in a geospatial context sets PostGIS in a category on its own.

PostGIS allows you to execute both attribute and spatial queries against a dataset. Recall the point from *Chapter 2*, *Geospatial Data*, that a typical spatial data set is comprised of multiple data types including geometry, attributes (one or more columns of data in a row), and in most cases, indexing data. In PostGIS, you can query attribute data as you would do to any database table using SQL. This capability is not surprising as attribute data is stored in a traditional database structure. However, you can also query geometry using SQL syntax. Spatial operations are available through SQL functions, which you include as part of queries. The following sample PostGIS SQL statement creates a 14.5 kilometer buffer around the state of Florida:

```
SELECT ST_Buffer(the_geom, 14500)
FROM usa_states
WHERE state = 'Florida'
```

The FROM clause designates the usa_states layer as the location of the query. We filter that layer by isolating Florida in the WHERE clause. Florida is a value in the column state of the usa_states layer. The SELECT clause performs the actual spatial selection on the geometry of Florida normally contained in the column the_geom using the PostGIS ST_Buffer() function. The column the_geom is the geometry column for the PostGIS layer in this instance. The ST in the function name stands for Spatial Type. The ST_Buffer() function accepts a column containing spatial geometries and a distance in the map units of the underlying layer. The map units in usa_states layer are expressed in meters, so 14.5 km would be 14,500 meters in the preceding example. Recall the point from *Chapter 1, Learning Geospatial Analysis with Python* that buffers like this query are used for proximity analysis. It just so happens that the State of Florida water boundary expands 9 nautical miles or approximately 14.5 km into the Gulf of Mexico from the state's western and northwestern coastlines.

The following image shows the official Florida state water boundary as a dotted line, which is labeled on the map:

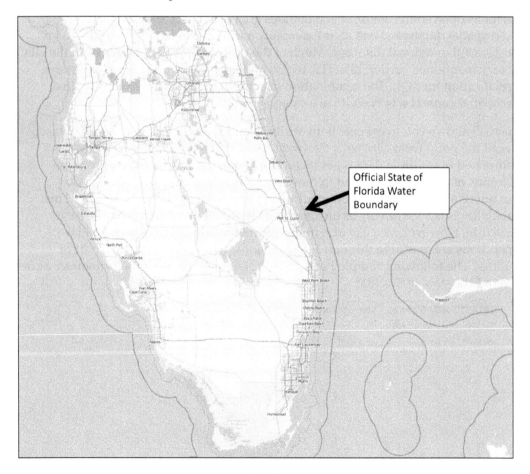

After applying the 9 nautical mile buffer, you can see in the following image that the result, highlighted in orange, is quite close to the official legal boundary based on detailed ground surveys:

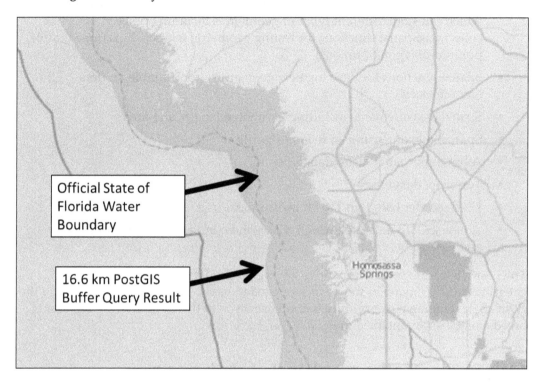

The website GISTutor.com has an excellent interactive online tutorial, which allows you to execute spatial queries against a continental U.S. map and see the result immediately on a web map. A solid written introductory PostGIS tutorial can be found at the following URL:

http://workshops.boundlessgeo.com/postgis-intro/

Currently, PostGIS maintains the following feature set:

- Geospatial geometry types including points, linestrings, polygons, multipoints, multilinestrings, multipolygons, and geometry collections, which can store different types of geometries including other collectionsSpatial functions for testing geometric relationships (for example, point-in-polygon or unions)

- Spatial functions for deriving new geometries (for example, buffers and intersects)

- Spatial measurements including perimeter, length, and area

- Spatial indexing using an R-Tree algorithm

- A basic geospatial raster data type

- Topology data types

- US Geocoder based on TIGER census data

- A new JSONB data type which allows indexing and querying of JSON and GeoJSON

The PostGIS feature set is competitive among all geodatabases and the most extensive among any open source or free geodatabases. The active momentum of the PostGIS development community is another reason of this system being the best of breed. PostGIS is maintained at `http://postgis.net`.

Other spatially-enabled databases

PostGIS is the gold standard among free and open source geospatial databases. However, there are several other systems that you should be aware of as a geospatial analyst. This list includes both commercial and open source systems with varying degrees of geospatial support.

Geodatabases have evolved in parallel to geospatial software, standards, and the Web. The Internet has driven the need for large, multiuser geospatial database servers able to serve large amounts of data. The following image, courtesy of `www.OSGeo.org`, shows how geospatial architectures have evolved with a significant portion of this evolution happening at the database level:

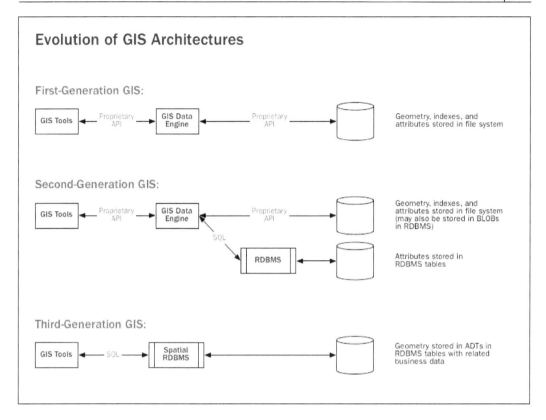

Oracle spatial and graph

The Oracle relational database is a widely used database system typically used by very large organizations because of its cost and large scalability. It is also extremely stable and fast. It runs some of the largest and most complicated databases in the world. It is often found in hospitals, banks, and government agencies managing millions of critical records.

Geospatial data capability first appeared at Oracle Version 4 as a modification by the **Canadian Hydrographic Service** (**CHS**). CHS also implemented Oracle's first spatial index in the form of an unusual but efficient three-dimensional helical spiral. Oracle subsequently incorporated the modification and released the Oracle **Spatial Database Option** (**SDO**) at Version 7 of the main database. The SDO system became Oracle Spatial at Oracle Version 8. The database schema of Oracle Spatial still has the SDO prefix on some column and table names similar to how PostGIS uses the OGC convention **ST** (**spatial type**) to separate spatial information from traditional relational database tables and functions at the schema level.

As of 2012, Oracle began calling the package Oracle Spatial and Graph to emphasize the network data module. This module is used for analyzing networked datasets such as transportation or utilities. However, the module can also be used against abstract networks such as social networks. The analysis of social network data is a common target for big data analysis, which is now a growing trend. Big data social network analysis is likely the reason Oracle changed the name of the product.

As a spatial, Oracle Spatial has the following capabilities:

- A geospatial data schema
- A spatial indexing system which is now based on an R-tree index
- An SQL API for performing geometric operations
- A spatial data tuning API to optimize a particular dataset
- A topology data model
- A network data model
- A GeoRaster data type to store, index, query, and retrieve raster data
- Three-dimensional data types including **Triangulated Irregular Networks (TINs)** and LIDAR point clouds
- A geocoder to search location names and return coordinates
- A routing engine for driving direction-type queries
- Open Geospatial Consortium-compliance

Oracle Spatial and PostGIS are reasonably comparable and are both commonly used. You will see these two systems sooner or later as data sources while performing geospatial analysis.

 Oracle Spatial and Graph is sold separately from Oracle itself. A little-known fact is that the SDO data type is native to the main Oracle database. If you have a simple application which just inputs points and retrieves them, you can use the main Oracle API to add, update, and retrieve SDOs without Oracle Spatial and Graph.

The U.S. **Bureau of Ocean Energy, Management, Regulation, and Enforcement (BOEMRE)** uses Oracle to manage environmental, business, and geospatial data for billions of dollars' worth of oil, gas, and mineral rights in one of the largest geospatial systems in the world. The following map is courtesy of BOEMRE:

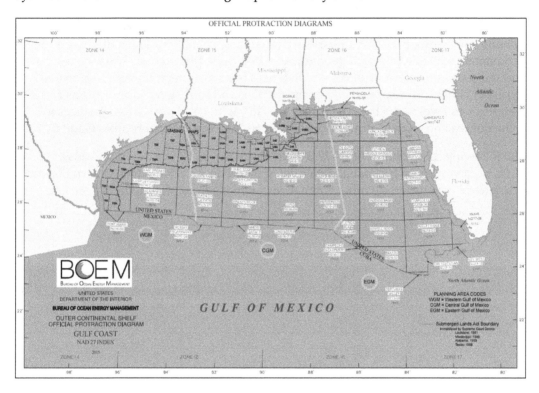

Oracle Spatial and Graph can be found online at the following URL: `http://www.oracle.com/us/products/database/options/spatial/overview.`

ArcSDE

ArcSDE is Esri's **Spatial Data Engine (SDE)**. It is now rolled into Esri's ArcGIS Server product after over a decade of being a standalone product. What makes ArcSDE interesting is that the engine is mostly database independent supporting multiple database backends. ArcSDE supports IBM DB2, Informix, Microsoft SQL Server, Oracle and PostgreSQL as data storage systems. While ArcSDE has the ability to create and manage a spatial schema from scratch on systems such as Microsoft SQL Server and Oracle, it uses native spatial engines if available. This arrangement is the case for IBM DB2, Oracle, and PostGreSQL. For Oracle, ArcSDE manages the table structure but can rely on the Oracle SDO data type for feature storage.

Like the previously mentioned geodatabases, ArcSDE also has a rich spatial selection API and can handle raster data. However, ArcSDE does not have as rich a SQL spatial API as Oracle and PostGIS. Esri technically supports basic SQL functionality related to ArcSDE, but it encourages users and developers to use Esri software or programming APIs to manipulate data stored through ArcSDE as it is designed to be a datasource for Esri software. Esri does provide software libraries for developers to build applications outside of Esri software using ArcSDE or Esri's file-based geodatabase called a personal geodatabase. But, these libraries are black boxes and the communication protocol ArcSDE uses has never been reverse engineered. Typically, interaction happens between ArcSDE and third-party applications at the web services level using the ArcGIS Server API, which supports OGC services to some degree and a fairly straightforward REST API service that returns GeoJSON.

The following screenshot is of the U.S. federal site `http://catalog.data.gov/dataset`, a very large geospatial data catalog based on ArcSDE, which in turn networks U.S. federal data including other ArcSDE installations from other federal agencies:

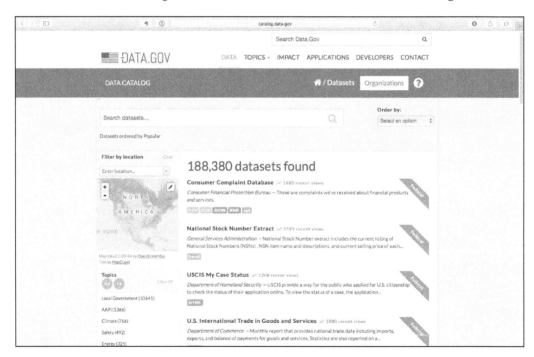

ArcSDE is integrated into ArcGIS Server; however, information on it remains at `http://www.esri.com/software/arcgis/arcsde`.

Microsoft SQL Server

Microsoft added spatial data support to its flagship database product in Microsoft SQL Server 2008. It has gradually improved since that version, but still is nowhere near as sophisticated as Oracle Spatial or PostGIS. Microsoft supports the same data types as PostGIS with slightly different naming conventions with the exception of rasters, which are not directly supported. It also supports output to WKT and WKB formats.

It offers some very basic support for spatial selection, but it is obviously not a priority for Microsoft at the moment. This limited support is likely the case because it is all that can be used for Microsoft software mapping components and several third party engines can provide spatial support on top of SQL Server.

Microsoft's support for spatial data in SQL Server is documented at the following link:

```
http://msdn.microsoft.com/en-us/library/bb933790.aspx
```

MySQL

MySQL, another highly popular free database, provides nearly the same support as Microsoft SQL Server. The OGC Geometry types are supported with basic spatial relationship functions. Through a series of buyouts, MySQL has become the property of Oracle. While Oracle currently remains committed to MySQL as an open source database, this purchase has brought the ultimate future of the world's most popular open source database into question. But as far as geospatial analysis is concerned, MySQL is barely a contender and unlikely to be the first choice for any project.

For more information on MySQL spatial support visit the following link:

```
http://dev.mysql.com/doc/refman/5.6/en/spatial-extensions.html
```

SpatiaLite

SpatiaLite is an extension for the open source SQLite database engine. SQLite uses a file database and is designed to be integrated into applications instead of the typical client server model used by most relational database servers. SQLite has spatial data types and spatial indexing already, but SpatiaLite adds support for the OGC Simple Features Specification as well as map projections.

SpatiaLite can be found at `http://www.gaia-gis.it/gaia-sins/`.

Routing

Routing is a very niche area of computational geometry. It is also a very rich field of study that goes far beyond the familiar driving directions use case. The requirements for a routing algorithm are simply a networked dataset and impedance values that affect the speed of travel on that network. Typically, the dataset is vector-based but raster data can also be used for certain applications. The two major contenders in this area are Esri's Network Analyst and the open source pgRouting engine for PostGIS. The most common routing problem is the most efficient way to visit a number of point locations. This problem is called the **travelling salesman problem** (**TSP**). The TSP is one of the most intensely studied problems in computational geometry. It is often considered the benchmark for any routing algorithm. More information on the TSP can be found at `http://en.wikipedia.org/wiki/Travelling_salesman_problem`.

Esri Network Analyst and Spatial Analyst

Esri's entry into the routing arena, Network Analyst, is a truly generic routing engine, which can tackle most routing applications regardless of context. Spatial Analyst is another Esri extension that is raster focused and can perform *least cost path* analysis on raster terrain data.

The ArcGIS Network Analyst product page is located on Esri's website at `http://www.esri.com/software/arcgis/extensions/networkanalyst`.

pgRouting

The **pgRouting** extension for PostGIS adds routing functionality to the geodatabase. It is oriented towards road networks but can be adapted to work with other types of networked data. The following image shows a driving distance radius calculation output by pgRouting and displayed in QGIS. The points are color-coded from green to red based on proximity to the starting location. As shown in the following diagram, the points are nodes in the network dataset, courtesy of `QGIS.org`, which in this case are roads:

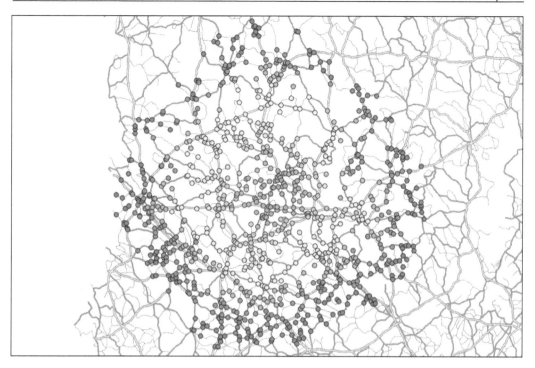

The pgRouting PostGIS extension is maintained at `http://pgrouting.org/`.

Desktop tools (including visualization)

Geospatial analysis requires the ability to visualize output in order to be complete. This fact makes tools, which can visualize data absolutely critical to the field. There are two categories of geospatial visualization tools. The first is geospatial viewers and the second is geospatial analysis software. The first category — geospatial viewers, allows you to access, query, and visualize data but not edit it in any way. The second category allows you to perform those items and edit data too. The main advantage of viewers is that they are typically lightweight pieces of software that launch and load data quickly. Geospatial analysis software requires far more resources to be able to edit complex geospatial data, so it loads slower and often renders data more slowly to provide dynamic editing functionality.

Quantum GIS

Quantum GIS, more commonly known as **QGIS**, is a complete open source geographic information system. QGIS falls well within the geospatial analysis category in the two categories of visualization software. The development of the system began in 2002 and version 1.0 was released in 2009.

It is the best showcase of most of the libraries mentioned earlier in this chapter. QGIS is written in C++ using the **Qt** library for the GUI. The GUI is well designed and easy to use. In fact, a geospatial analyst trained on a proprietary package like Esri's ArcGIS or Manifold System will be right at home using QGIS. The tools and menu system are logical and typical of a GIS system. The overall speed of QGIS is as good as or better than any other system available.

A nice feature of QGIS is that the underlying libraries and utility programs are just below the surface. Modules can be written by any third party in Python and added to the system. QGIS also has a robust online package management system to search for, install, and update these extensions. The Python integration includes a console that allows you to issue commands at the console and see the results in the GUI. QGIS isn't the only software to offer this capability.

Like most geospatial software packages, with Python integration it installs a complete version of Python if you use the automated installer. There's no reason to worry if you already have Python installed. Having multiple versions of Python on a single machine is fairly common and well supported. Many people have multiple versions of Python on their computers for testing software or because it is such a common scripting environment for so many different software packages. When the Python console is running in QGIS, the entire program API is available through an automatically loaded object called `qgis.utils.iface`. The following screenshot shows QGIS with the Python console running:

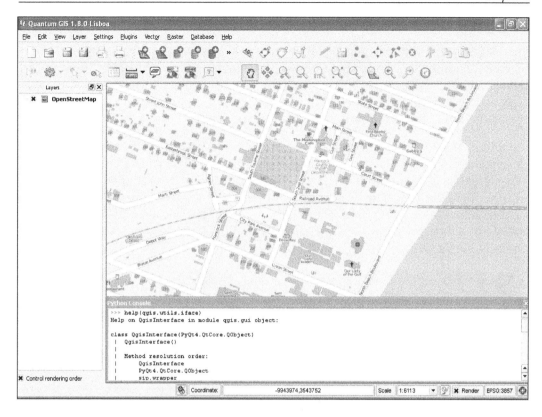

Because QIS is based on GDAL/OGR and GEOS, and it can use PostGIS, it supports all of the data sources offered by those packages. It also has nice raster processing features too. QGIS works well for producing paper maps or entire map books using available extensions.

QGIS is well documented through the QGIS website at the following link: `http://www.qgis.org/en/documentation.html`

You can also find numerous online and video tutorials by searching for QGIS and a particular operation.

OpenEV

OpenEV is an open source geospatial viewer originally developed by Atlantis Scientific around 2002, which became Vexcel before a buyout by Microsoft. Vexcel developed OpenEV as a freely downloadable satellite image viewer for the Canadian Geospatial Data Infrastructure. It is built using GDAL and Python and is partially maintained by GDAL-creator Frank Warmerdam.

OpenEV is one of the fastest raster viewers available. Despite being originally designed as a viewer, OpenEV offers all of the utility of GDAL/OGR and PROJ.4. While created as a raster tool, it can overlay vector data such as shapefiles and even supports basic editing. Raster images can also be altered using the built-in raster calculator, and data formats can be converted, reprojected, and clipped. The following screenshot shows a 25 megabyte, 16-bit integer geotiff elevation file in an OpenEV viewer window:

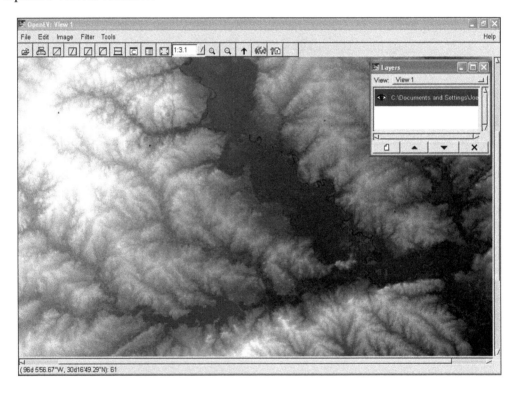

OpenEV is built largely in Python and offers a Python console with access to the full capability of the program. The OpenEV GUI isn't as sophisticated as other tools like QGIS. For example, you cannot drag and drop geospatial datasets into the viewer like you can do in QGIS. But, the raw speed of OpenEV makes it very attractive for simple raster viewing or basic processing and data conversion.

The OpenEV homepage is available at `http://openev.sourceforge.net/`.

GRASS GIS

The GRASS is one of the oldest continuously developed geospatial systems in existence. The U.S. Army Corps of Engineers began GRASS development in 1982. It was originally designed to run on UNIX systems. In 1995, the Army released the last patch, and the software was transferred to community development where it has remained ever since.

Even though the user interface was redesigned, GRASS still feels somewhat esoteric to modern GIS users. However, because of its decades old legacy and non-existent price tag, many geospatial workflows and highly specialized modules have been implemented in GRASS over the years, making it highly relevant to many organizations and individuals especially in research communities. For these reasons, GRASS is still actively developed.

GRASS has also been integrated with QGIS, so the more modern and familiar QGIS GUI can be used to run GRASS functions. GRASS is also deeply integrated with Python and can be used as a library or command-line tool. The following screenshot shows some landform analysis in the native GRASS GUI built using the `wxPython` library:

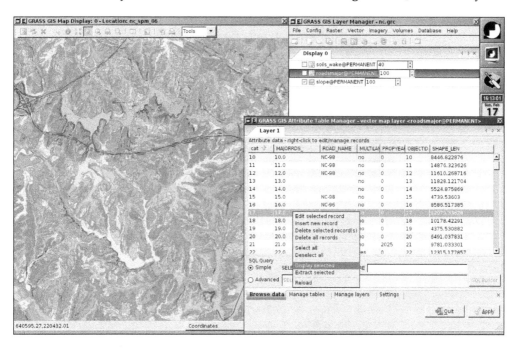

GRASS is housed online at `http://grass.osgeo.org/`.

uDig

The program uDig is a Java-based GIS viewer. It is built on top of the Eclipse platform originally created by IBM. Eclipse is designed as an **integrated development environment** (IDE) for programmers. But, there are a lot of interface similarities between IDE and a GIS, so the modification works quite well. The following screenshot shows the core Eclipse IDE platform as typically used by programmers:

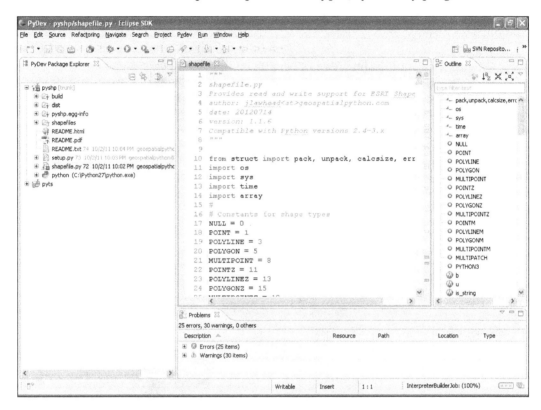

The following screenshot demonstrates the uDig GIS built on top of Eclipse:

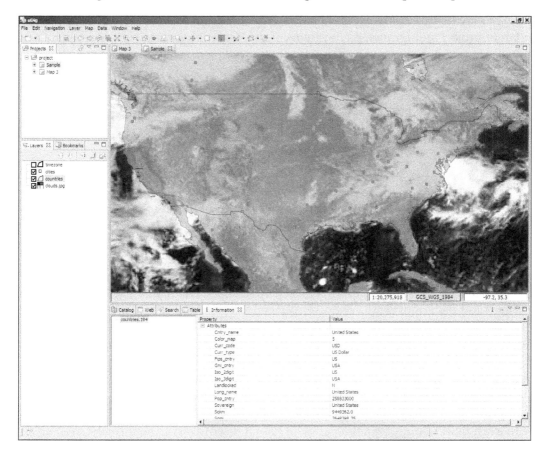

uDig is designed primarily as a thick-client viewer for web services (for example, WMS and WFS) and common data types. Because of the Eclipse platform, the developers encourage third-party plugins or even full-blown application development on top of uDig. The program does support more advanced analysis by using the GRASS GIS program and its Java bindings called JGRASS.

The uDig homepage is located at `http://udig.refractions.net/`.

gvSIG

Another Java-based desktop GIS is gvSIG. The gvSIG project began in 2004 as part of a larger project to migrate the IT systems of the Regional Ministry of Infrastructure and Transport of Valencia, Spain to free software. The result was gvSIG, which has continued to mature. The feature set is mostly comparable to QGIS with some unique capabilities as well. The official gvSIG project has a very active fork called **gvSIG Community Edition** (**gvSIG CE**). There is also a mobile version called gvSIG mobile. The gvSIG code base is open source. The official homepage for gvSIG is available at `http://www.gvsig.org/web/`.

OpenJUMP

OpenJUMP is another open source Java-based Desktop GIS. **JUMP** stands for **Java Unified Mapping Platform** and was originally created by Vivid Solutions for the Government of British Columbia. After Vivid Solutions delivered JUMP, development stopped. Vivid Solutions eventually released JUMP to the open source community where it was renamed OpenJUMP. OpenJUMP has the ability to read and write shapefiles, OGC GML, and supports PostGIS databases. It can also display some image formats and data from OGC WMS and WFS services. It has a plugin architecture and can also serve as a development platform for custom applications. You can find out more about OpenJUMP on the official web page at `http://www.openjump.org/`.

Google Earth

Google Earth is so ubiquitous that it hardly seems worth mentioning. But as you learn more about geospatial analysis, you'll discover that there is a lot of misinformation surrounding Google Earth. The first release of EarthViewer 3D came in 2001 and was created by a company called Keyhole Inc. and the EarthViewer 3D project were funded by the non-profit venture capital firm In-Q-Tel, which in turn is funded by the U.S. Central Intelligence agency. This cloak-and-dagger spy agency lineage and the subsequent purchase of Keyhole by Google to create and distribute Google Earth, brought global attention to the geospatial analysis field.

Since the first release of the software as Google Earth in 2005, Google has continually refined the software. Some of the notable additions are creating Google Moon, Google Mars, Google Sky, and Google Oceans. These are virtual globe applications, which feature data from the Moon and Mars with the exception of Google Oceans, which adds sea-floor elevation mapping known as bathymetry to Google Earth. Google also released a Google Earth browser plugin allowing a simplified version of the globe in a browser. The plugin has a JavaScript API, which allows reasonably sophisticated control over the data and position of the view for the plugin. However this API is ending in December, 2015.

Google Earth introduced the idea of the spinning virtual globe concept for exploration of geographic data. After centuries of looking at 2D maps or low-resolution physical globes, flying around the Earth virtually and dropping into a street corner anywhere in the world was mind-blowing—especially for geospatial analysts and other geography enthusiasts, as depicted in the following screenshot of Google Earth overlooking the New Orleans, Louisiana Central Business District:

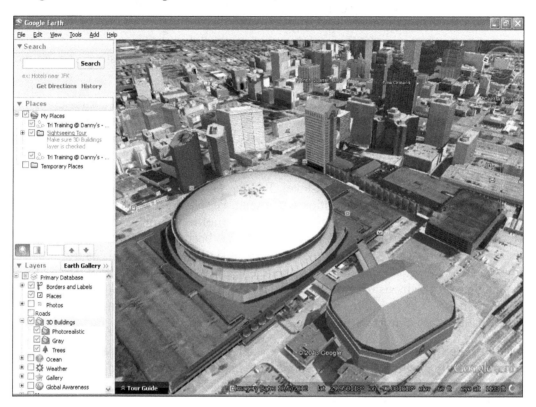

Just as Google had revolutionized web mapping with its tile-based *slippy* mapping approach, the virtual globe concept was a major boost to geospatial visualization.

After the initial excitement wore off, many geospatial analysts realized that Google Earth was a very animated and fun geographic exploration tool, but it really had very limited utility for any kind of meaningful geospatial analysis. Google Earth falls squarely into the realm of geospatial viewer software. The only data format it consumes is its native **Keyhole Markup Language** (**KML**), which is an all-in-one data and styling format discussed in *Chapter 2, Geospatial Data*. As this format is now an OGC standard, consuming only one data format immediately limits the utility of any tool. Any project involving Google Earth must first begin with complete data conversion and styling in KML, reminiscent of geospatial analysis from around 10-20 years ago. The tools which do support KML, including Google Maps, support a limited subset of KML.

Google Earth's native dataset has global coverage, but it is a mixture of datasets spanning several years and sources. Google has greatly improved the inline metadata in the tool, which identifies the source and approximate date of the current view. But this method creates confusion among lay people. Many people believe that the data in Google Earth is updated far more frequently than it really is. The Google Street View system, showing street-level, 360-degree views of much of the world, has helped correct this misperception somewhat. People are able to easily identify images of familiar locations as several years old. Another common misperception created by Google Earth is that the entire world has been mapped in detail and therefore creating a base map for geospatial analysis should be trivial. As discussed in *Chapter 2, Geospatial Data*, mapping an area of interest is far easier than it used to be a few years ago by using modern data and software, but it is still a complex and labor intensive endeavor. This misperception is one of the first customer expectations a geospatial analyst must manage while starting a project.

Despite these misperceptions, the impact Google has had on geospatial analysis is almost entirely positive. For decades, one of the most difficult challenges to growing the geospatial industry was to convince potential stakeholders that geospatial analysis is almost always the best approach while making decisions about people, resources, and the environment. This hurdle stands in sharp contrast to a car dealer. When a potential customer comes to a car lot, the salesman doesn't have to convince the buyer that they need a car but just about the type of car. Geospatial analysts have to first educate project sponsors on the technology and then convince them that the geospatial approach was the best way to address a challenge. Google has largely eliminated those steps for the analysts.

Google Earth can be found online at `http://www.google.com/earth/index.html`.

NASA World Wind

NASA World Wind is an open source, virtual globe, and geospatial viewer, originally released by the U.S. **National Aeronautics and Space Administration (NASA)** in 2004. It was originally based on Microsoft's .NET framework making it a Windows-centric application. The following screenshot of NASA World Wind looks similar to Google Earth:

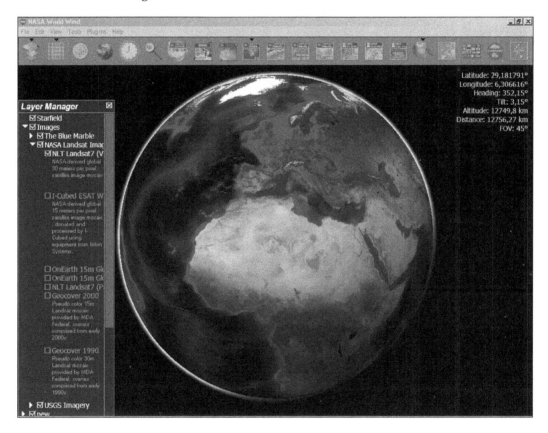

In 2007, a Java-based **software development kit (SDK)** was released called World Wind Java, which made World Wind more cross-platform. The transition to Java also led to the creation of a browser plugin for World Wind.

The World Wind Java SDK is considered an SDK and not a desktop application like the .NET version. However, the demos included with the SDK provide a viewer without any additional development. While NASA World Wind was originally inspired by Google Earth, its status as an open source project takes it in an entirely different direction. Google Earth is a generalist tool bounded by the limits of the KML specification. NASA World Wind is now a platform upon which anyone can develop without limits. As new types of data become available and computing resources grow, the potential of the virtual globe paradigm certainly holds more potential for geospatial visualization, which has not been explored yet.

NASA World Wind is available online at `http://worldwind.arc.nasa.gov/java/`.

ArcGIS

Esri walks the line of one of the greatest promoters of the geospatial analytical approach to understanding our world and a privately held, profit-making business, which must look out for its own interests to a certain degree. The ArcGIS software suite represents every type of geospatial visualization known including vector, raster, globes, and 3D. It is also a market leader in many countries. As described in the geospatial software map earlier in this chapter, Esri has increasingly incorporated open source software into its suite of tools including GDAL for raster display and Python as the scripting language for ArcGIS.

The following screenshot shows the core ArcGIS application ArcMap with marine tracking density data analysis. The interface shares a lot in common with QGIS as shown in this image courtesy of `MarineCadastre.gov`:

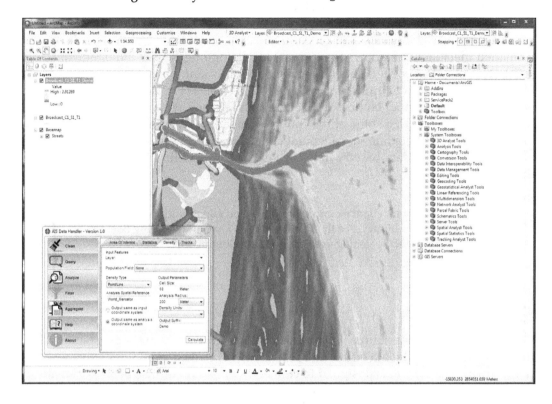

The ArcGIS product page is available online at `http://www.esri.com/software/ arcgis`.

Metadata management

Internet distribution of data has increased the importance of metadata. Data custodians are able to release a dataset to the entire world for download without any personal interaction. The metadata record of a geospatial dataset can follow that to help ensure that the integrity and accountability for that data is maintained. Properly formatted metadata also allows for automated cataloguing, search indexing, and integration of datasets. Metadata has become so important that a common mantra within the geospatial community is *Data without metadata isn't data*, meaning that a geospatial dataset cannot be fully utilized and understood without metadata. The following section will list some of the common metadata tools that are available. The OGC standard for metadata management is the **Catalog Service for the Web (CSW)**, which creates a metadata-based catalog system and an API for distributing and discovering datasets.

> For an excellent example of a CSW and client built using Python's
> `pycsw` library, see the **Pacific Islands Ocean Observing System
> (PacIOOS)** catalog available at the following link:
>
> `http://pacioos.org/search/`

GeoNetwork

GeoNetwork is an open source, Java-based catalog server to manage geospatial data. It includes a metadata editor and search engine as well as an interactive web map viewer. The system is designed to connect spatial data infrastructures globally. It can publish metadata through the web using the metadata editing tools. It can publish the spatial data as well through the embedded GeoServer map server. It has user and group security permissions and web and desktop configuration utilities. GeoNetwork can also be configured to harvest metadata from other catalogs at scheduled intervals. The following screenshot is of the United Nations Food and Agriculture Organization's implementation of the GeoNetwork:

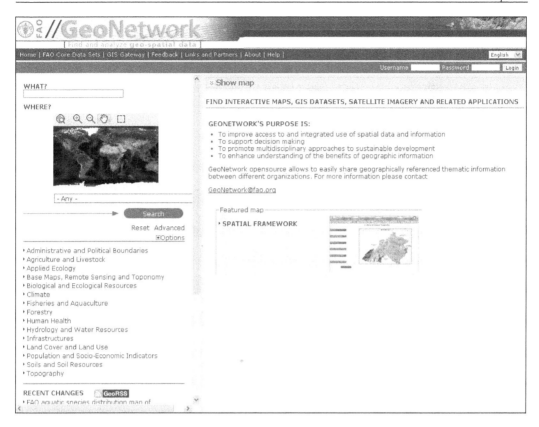

You can find out more about GeoNetwork at http://geonetwork-opensource.org/.

CatMDEdit

CatMDEdit is another Java-based metadata editor focused on the geospatial data from the National Geographic Institute of Spain and several collaborators. CatMDEdit can exchange metadata records using XML and **Resource Description Framework (RDF)** standards and style, and then transform the metadata for viewing in different formats; it also enables visualization of geospatial data, integration with gvSIG and has many other features. The CatMDEdit website is located at http://catmdedit. sourceforge.net/.

Summary

In this chapter, you learned the hierarchy of geospatial analysis software. You learned a framework for approaching the hundreds of existing geospatial software packages and libraries by categorizing them into one or more major functions including data access, computational geometry, raster processing, visualization, and metadata management.

We also examined the commonly used foundation libraries including GDAL, OGR, PROJ.4, and GEOS found again and again in geospatial software. You can approach any new piece of geospatial software, trace it back to these core libraries and then ask, *What is the value added?* to better understand the package. If the software isn't using one of these libraries, you need to ask, *Why are these developers going against the grain?* to understand what that system brings to the table.

Python was only mentioned a few times in this chapter to avoid any distraction in understanding the geospatial software landscape. But, as we will see, Python is interwoven into every single piece of software in this chapter and is a fully capable geospatial tool in its own right. It is no coincidence that Python is the official scripting language of ArcGIS, QGIS, GRASS, and many other packages. It is also not by chance that GDAL, OGR, PROJ.4, CGAL, JTS, GEOS, and PostGIS all have Python bindings. And as for the packages not mentioned here, they are all within Python's grasp as well through the Jython Java distribution, the IronPython .NET distribution, Python's various database and Web APIs, and the built-in **ctypes** module. As a geospatial analyst, if there's one technology you can't afford to pass up, it's Python.

4

Geospatial Python Toolbox

The first three chapters of this book covered the history of geospatial analysis, the types of geospatial data used by analysts, and the major software and libraries found within the geospatial industry. We used some simple Python examples here and there to illustrate certain points, but we focused mainly on the field of geospatial analysis independent of any specific technology.

Starting here, we will be using Python to conquer geospatial analysis and we will continue with that approach for the rest of the book. In this chapter, we'll discover the Python libraries used to access the different types of data found in the Vector data and Raster data sections of *Chapter 2*, *Geospatial Data*. Some of these libraries are pure Python and some are bindings to the different software packages found in *Chapter 3*, *The Geospatial Technology Landscape*.

We will examine pure Python solutions whenever possible. Python is a very capable programming language, but some operations, particularly in remote sensing, are too computationally intensive and therefore impractical using pure Python or other interpreted languages. Fortunately, nearly every aspect of geospatial analysis can be addressed in some way through Python even if it is binding to a highly efficient C, C++, or other compiled-language library.

We will avoid using broad scientific libraries which cover other domains beyond geospatial analysis to keep solutions as simple as possible. There are many reasons to use Python for geospatial analysis, but one of the strongest arguments is its portability. Python is a ubiquitous programming language officially available as a compiled installation on over 20 platforms according to the python.org website. It comes as standard with most Linux distributions and is available on most major smart phone operating systems as well. The Python source distribution usually compiles on any platform supporting C.

Furthermore, Python has been ported to Java as the **Jython** distribution and to the .NET **Common Language Runtime (CLR)** as **IronPython**. Python also has versions such as Stackless Python for massively concurrent programs. There are versions of Python designed too to run on cluster computers for distributed processing. Python is also available on many hosted application servers that do not allow you to install custom executables such as the Google App Engine platform, which has a Python API. Modules written in pure Python using the standard library will almost always run on any of the platforms that we just mentioned.

Each time you add a third-party module which relies on bindings to external libraries in other languages, you reduce Python's inherent portability. You also add a layer of complexity to fundamentally change the code by adding another language to the mix. Pure Python keeps things simple. Also Python bindings to external libraries tend to be automatically or semi-automatically generated. These automatically generated bindings are very generic and esoteric, and they simply connect Python to a C or C++ API using the method names from that API instead of following the best practices for Python. There are, of course, notable exceptions to this approach driven by project requirements which may include speed, unique library features, or frequently updated libraries where an automatically generated interface is preferable.

Installing third-party Python modules

We'll make a distinction between modules which are included as part of Python's standard library and modules which must be installed. In Python, the words module and library are used interchangeably. To install libraries, you either get them from the **Python Package Index (PyPI)** or in the case of a lot of geospatial modules, you download a specialized installer. PyPI acts as the official software repository for libraries and offers some easy-to-use setup programs, which simplify installing packages. You can use the `easy_install` program, which is especially good on Windows, or the pip program more commonly found on Linux and Unix systems. Once it's installed, you can then install third-party packages simply by running the following code:

```
easy_install <package name>
```

For installing `pip`, you run the following code:

```
pip install <package name>
```

Links will be provided to installers and instructions for packages not available on PyPI. You can manually install third-party Python modules by downloading the Python source code and putting it in your current working directory, or you can put it in your Python site-packages directory. These two directories are available in Python's search path when you try to import a module. If you put a module in your current working directory, it'll only be available when you start Python from that directory.

If you put it in your site-packages directory, it'll be available every time you start Python. The site-packages directory is specifically meant for third-party modules. To locate the site-packages directory for your installation, you ask Python's sys module. The sys module has a path attribute that is a list of all directories in Python's search path. The site-packages directory should be the last one which you can locate by specifying an index of -1, as shown in the following code:

```
>>> import sys
>>> sys.path[-1]
'C:\\Python34\\lib\\site-packages'
```

If that call doesn't return the site-packages path, just look at the entire list to locate it, as shown in the following code:

```
>>> sys.path
['', 'C:\\WINDOWS\\system32\\python34.zip', 'C:\\Python34\\DLLs',
  'C:\\Python34\\lib', 'C:\\Python34\\lib\\plat-win
', 'C:\\Python34\\lib\\lib-tk', 'C:\\Python34',
  'C:\\Python34\\lib\\site-packages']
```

These installation methods will be used for the rest of the book. You can find the latest Python version, source code for your platform installation, and compilation instructions at http://python.org/download/.

The Python virtualenv module allows you to easily create an isolated copy of Python for a specific project without affecting your main Python installation or other projects. Using this module, you can have different projects with different versions of the same library. Once you have a working code base, you can then keep it isolated from changes to the modules you used or even Python itself. The virtualenv module is simple to use and can be used for any example in this book; however, explicit instructions on its use are not included. To get started with virtualenv, follow this simple guide:

http://docs.python-guide.org/en/latest/dev/virtualenvs/

Installing GDAL

The **Geospatial Data Abstraction Library** (**GDAL**), which includes OGR, is critical to many of the examples in this book and is also one of the more complicated Python setups as well. For these reasons, we'll discuss it separately here. The latest GDAL bindings are available on PyPI, however the installation requires a few more steps because of additional resources needed by the GDAL library.

There are three ways to install GDAL for use with Python:

- Compile it from source code
- Install it as part of a larger software package
- Install a binary distribution and then Python bindings

If you have experience with compiling C libraries as well as the required compiler software, then the first option gives you the most control. However, it is not recommended if you just want to get going as quickly as possible because even experienced software developers can find compiling GDAL and the associated Python bindings challenging. Instructions for compiling GDAL on leading platforms can be found at `http://trac.osgeo.org/gdal/wiki/BuildHints`. There are also basic build instructions on the PyPI GDAL page. Have a look at `https://pypi.python.org/pypi/GDAL`.

The second option is by far the quickest and the easiest. The **Open Source Geospatial Foundation** (**OSGeo**) distributes an installer called **OSGeo4W** which installs all of the top open source geospatial packages on Windows at the click of a button. If you are on Linux, there is another package with distributions for both Linux and Windows called **FWTools**. OSGeo4W can be found at `http://trac.osgeo.org/osgeo4w/`.

FWTools is available online at `http://fwtools.maptools.org/`.

While these packages are the easiest to work with, they come with their own version of Python. If you already have Python installed, then having another Python distribution just to use certain libraries may be problematic. In that case, the third option may be for you.

The third option installs a pre-compiled binary specific to your Python version. This method is the best compromise between ease of installation and customization. The catch is you must make sure the binary distributions and the corresponding Python bindings are compatible with each other, your Python version, and in many cases your operating system configuration.

Windows

The installation of GDAL for Python on Windows becomes easier and easier each year. To install GDAL on Windows, you must first check whether you are running the 32-bit or 64-bit version of Python. To do so, just start your Python interpreter at a command prompt, as shown in the following code:

```
Python 3.4.2 (v3.4.2:ab2c023a9432, Oct 6 2014, 22:15:05) [MSC v.1600
  32 bit (Intel)] on win32
Type "help", "copyright", "credits" or "license" for more
  information.
```

So, based on this instance, we see Python is version 3.4.2 for `win32`, which means it is the 32-bit version. Once you have this information, go to the following URL:

```
http://www.lfd.uci.edu/~gohlke/pythonlibs/#gdal
```

This web page contains Python Windows binaries and bindings for nearly every open source scientific library. On that web page, in the **GDAL** section, find the release that matches your version of Python. The release names use the abbreviation `cp` for C Python followed by the major Python version number and either `win32` for 32-bit Windows or `win_amd64` for 64-bit Windows. In the previous example, we would download the file named `GDAL-1.11.3-cp34-none-win32.whl`.

This download package is the newer Python pip **wheel** format. To install it, simply open a command prompt and type the following code:

```
pip install GDAL-1.11.3-cp34-none-win32.whl
```

Once the package is installed, open your Python interpreter and run the following commands to verify that GDAL is installed by checking its version:

```
Python 3.4.2 (v3.4.2:ab2c023a9432, Oct 6 2014, 22:15:05) [MSC v.1600 32
bit (Intel)] on win32
Type "help", "copyright", "credits" or "license" for more information.
>>> from osgeo import gdal
>>> gdal.__version__
1.11.3
```

GDAL should return its version as `1.11.3`.

> If you have trouble installing modules using `easy_install` or `pip` and PyPI, try to download and install the wheel package from the same site as the GDAL example.

Linux

GDAL installation on Linux varies widely by distribution. The following `gdal.org` binaries web page lists installation instructions for several distributions:

`http://trac.osgeo.org/gdal/wiki/DownloadingGdalBinaries`

Typically, your package manager will install both GDAL and Python bindings. For example, on Ubuntu, to install GDAL you run the following code:

```
sudo apt-get install gdal-bin
```

Then to install the Python bindings, you run the following command:

```
sudo apt-get install python-gdal
```

Also, most Linux distributions are set up to compile software already and their instructions are much simpler than those on Windows. Depending on the installation, you may have to import `gdal` and `ogr` as part of the `osgeo` package as shown in the following command:

```
>>> from osgeo import gdal
>>> from osgeo import ogr
```

Mac OS X

To install GDAL on OS X, you can also use the **Homebrew** package management system available at `http://brew.sh/`.

Alternatively, you can use the **MacPorts** package management system available at `https://www.macports.org/`.

Both of these systems are well documented and contain GDAL packages for Python 3. You really only need them for libraries which require a properly compiled binary written in C which has a lot of dependencies and includes many of the scientific and geospatial libraries.

Python networking libraries for acquiring data

The vast majority of geospatial data sharing is accomplished via the Internet. And Python is well equipped when it comes to networking libraries for almost any protocol. Automated data downloads are often an important step in automating a geospatial process. Data is typically retrieved from a website **Uniform Resource Locator (URL)** or a **File Transfer Protocol (FTP)** server. And because geospatial datasets often contain multiple files, data is often distributed as ZIP files.

A nice feature of Python is its concept of a file-like object. Most Python libraries which read and write data use a standard set of methods which allow you to access data from all different types of resources as if you were writing a simple file on disk. The networking modules in the Python standard library use this convention as well. The benefit of this approach is that it allows you to pass file-like objects to other libraries and methods, which recognize the convention without a lot of setup for different types of data distributed in different ways.

The Python urllib module

The Python `urllib` package is designed for simple access to any file with a URL address. The `urllib` package in Python 3 consists of several modules which handle different parts of managing web requests and responses. These modules implement some of Python's file-like object conventions starting with its `open()` method. When you call `open()`, it prepares a connection to the resource but does not access any data. Sometimes, you just want to grab a file and save it to the disk instead of accessing it in memory. This function is available through the `urllib.request.retrieve()` method.

The following example uses the `urllib.request.retrieve()` method to download the zipped shapefile named `hancock.zip`, which is used in other examples. We define the URL and the local file name as variables. The URL is passed as an argument as well as the filename we want to use to save it to our local machine which in this case is just `hancock.zip`:

```
>>> import urllib.request
>>> import urllib.parse
>>> import urllib.error
>>> url = "https://github.com/GeospatialPython/
   Learn/raw/master/hancock.zip"
>>> fileName = "hancock.zip"
>>> urllib.request.urlretrieve(url, fileName)
('hancock.zip', <httplib.HTTPMessage instance at 0x00CAD378>)
```

The message from the underlying `httplib` module confirms that the file was downloaded to the current directory. The URL and filename could have been passed to the `retrieve()` method directly as strings as well. If you specify just the filename, the download saves to the current working directory. You can also specify a fully qualified pathname to save it somewhere else. You can also specify a callback function as a third argument which will receive download status information for the file, so you can create a simple download status indicator or perform some other action.

The `urllib.request.urlopen()` method allows you to access an online resource with more precision and control. As mentioned previously, it implements most of the Python file-like object methods with the exception of the `seek()` method, which allows you to jump to arbitrary locations within a file. You can read a file online one line at a time, read all lines as a list, read a specified number of bytes, or iterate through each line of the file. All of these functions are performed in memory, so you don't have to store the data on disk. This ability is useful for accessing frequently updated data online, which you may want to process without saving to disk.

In the following example, we demonstrate this concept by accessing the **United States Geological Survey** (**USGS**) earthquake feed to view all of the earthquakes in the world, which have occurred within the last hour. This data is distributed as a **Comma-Separated Values** (**CSV**) file, which we can read line by line like a text file. CSV files are similar to spreadsheets and can be opened in a text editor or spreadsheet program. First, you'll open the URL and read the header with the column names in the file, and then you'll read the first line containing a record of a recent earthquake, as shown in the following lines of code:

```
>>> url = "http://earthquake.usgs.gov/earthquakes/feed/v1.0/
   summary/all_hour.csv"
>>> earthquakes = urllib.request.urlopen(url)
>>> earthquakes.readline()
'time,latitude,longitude,depth,mag,magType,nst,gap,dmin,rms,net,
   id,updated,place
\n'
>>> earthquakes.readline()
'2013-06-14T14:37:57.000Z,64.8405,-147.6478,13.1,0.6,Ml,
   6,180,0.09701805,0.2,ak,
ak10739050,2013-06-14T14:39:09.442Z,"3km E of Fairbanks,
   Alaska"\n'
```

We can also iterate through this file which is a memory efficient way to read through large files. If you are running this example in the Python interpreter as shown next, you will need to press the *Enter* or *Return* key twice to execute the loop. This action is necessary because it signals to the interpreter that you are done building the loop. In the following example, we abbreviate the output:

```
>>> for record in earthquakes: print(record)
2013-06-14T14:30:40.000Z,62.0828,-145.2995,22.5,1.6,
  Ml,8,108,0.08174669,0.86,ak,
ak10739046,2013-06-14T14:37:02.318Z,"13km ESE of Glennallen,
  Alaska"
...
2013-06-14T13:42:46.300Z,38.8162,-122.8148,3.5,0.6,
  Md,,126,0.00898315,0.07,nc,nc
72008115,2013-06-14T13:53:11.592Z,"6km NW of The Geysers,
  California"
```

FTP

FTP allows you to browse an online directory and download data using FTP client software. Until around 2004, when geospatial web services became very common, FTP was one of the most common ways to distribute geospatial data. FTP is less common now, but you occasionally encounter it when you're searching for data. Once again Python's **batteries included** standard library has a reasonable FTP module called `ftplib` with a main class called `FTP()`.

In the following example, we will access an FTP server hosted by the U.S. **National Oceanic and Atmospheric Administration (NOAA)** to access a text file containing data from the **Deep-ocean Assessment and Reporting of Tsunamis (DART)** buoy network used to watch for tsunamis around the world. This particular buoy is off the coast of Peru. We'll define the server and the directory path. Then we will access the server. All FTP servers require a username and password. Most public servers have a user called *anonymous* with the password *anonymous* as this one does. Using Python's `ftplib`, you can just call the `login()` method without any arguments to login as the default anonymous user. Otherwise, you can add the username and password as string arguments. Once we're logged in, we'll change to the directory containing the DART datafile. To download the file, we open up a local file called out and pass its `write()` method as a callback function to the `ftplib.ftp.retrbinary()` method, which simultaneously downloads the file and writes it to our local file.

Once the file is downloaded, we can close it to save it. Then we'll read the file and look for the line containing the latitude and longitude of the buoy to make sure that the data was downloaded successfully, as shown in the following lines of code:

```
>>> import ftplib
>>> server = "ftp.ngdc.noaa.gov"
>>> dir = "hazards/DART/20070815_peru"
>>> fileName = "21415_from_20070727_08_55_15_tides.txt"
>>> ftp = ftplib.FTP(server)
>>> ftp.login()
'230 Login successful.'
>>> ftp.cwd(dir)
'250 Directory successfully changed.'
>>> out = open(fileName, "wb")
>>> ftp.retrbinary("RETR " + fileName, out.write)
'226 Transfer complete.'
>>> out.close()
>>> dart = open(fileName)
>>> for line in dart:
...     if "LAT," in line:
...             print(line)
...             break
...
LAT,    LON     50.1663    171.8360
```

In this example, we opened the local file in binary write mode, and we used the `retrbinary()` ftplib method as opposed to `retrlines()`, which uses ASCII mode. Binary mode works for both ASCII and binary files, so it's always a safer bet. In fact, in Python, the binary read and write modes for a file are only required on Windows.

If you are just downloading a simple file from an FTP server, many FTP servers have a web interface as well. In that case, you could use `urllib` to read the file. FTP URLs use the following format to access data:

```
ftp://username:password@server/directory/file
```

This format is insecure for password-protected directories because you are transmitting your login information over the Internet. But for anonymous FTP servers, there is no additional security risk. To demonstrate this, the following example accesses the same file that we just saw but by using `urllib` instead of `ftplib`:

```
>>> dart = urllib.request.urlopen("ftp://" + server + "/" + dir +
  "/"   + fileName)
>>> for line in dart:
...     line = str(line, encoding="utf8")
...     if "LAT," in line:
...             print(line)
```

```
...          break
...
LAT,    LON     50.1663     171.8360
```

ZIP and TAR files

Geospatial datasets often consist of multiple files. For this reason, they are often distributed as ZIP or TAR file archives. These formats can also compress data, but their ability to bundle multiple files is the primary reason they are used for geospatial data. While the TAR format doesn't contain a compression algorithm, it incorporates the `gzip` compression and offers it as a program option. Python has standard modules for reading and writing both ZIP and TAR archives. These modules are called `zipfile` and `tarfile` respectively.

The following example extracts the `hancock.shp`, `hancock.shx`, and `hancock.dbf` files contained in the `hancock.zip` file we downloaded using `urllib` for use in the previous examples. This example assumes that the ZIP file is in the current directory:

```
>>> import zipfile
>>> zip = open("hancock.zip", "rb")
>>> zipShape = zipfile.ZipFile(zip)
>>> shpName, shxName, dbfName = zipShape.namelist()
>>> shpFile = open(shpName, "wb")
>>> shxFile = open(shxName, "wb")
>>> dbfFile = open(dbfName, "wb")
>>> shpFile.write(zipShape.read(shpName))
>>> shxFile.write(zipShape.read(shxName))
>>> dbfFile.write(zipShape.read(dbfName))
>>> shpFile.close()
>>> shxFile.close()
>>> dbfFile.close()
```

This example is more verbose than necessary for clarity. We can shorten this example and make it more robust by using a `for` loop around the `zipfile.namelist()` method without explicitly defining the different files as variables. This method is a more flexible and Pythonic approach, which could be used on ZIP archives with unknown contents, as shown in the following lines of code:

```
>>> import zipfile
>>> zip = open("hancock.zip", "rb")
>>> zipShape = zipfile.ZipFile(zip)
>>> for fileName in zipShape.namelist():
...     out = open(fileName, "wb")
...     out.write(zipShape.read(fileName))
...     out.close()
>>>
```

Now that you understand the basics of the `zipfile` module, let's take the files we just unzipped and create a TAR archive with them. In this example, when we open the TAR archive for writing, we specify the write mode as `w:gz` for gzipped compression. We also specify the file extension as `tar.gz` to reflect this mode, as shown in the following lines of code:

```
>>> import tarfile
>>> tar = tarfile.open("hancock.tar.gz", "w:gz")
>>> tar.add("hancock.shp")
>>> tar.add("hancock.shx")
>>> tar.add("hancock.dbf")
>>> tar.close()
```

We can extract the files using the simple `tarfile.extractall()` method. First, we open the file using the `tarfile.open()` method, and then extract it, as shown in the following lines of code:

```
>>> tar = tarfile.open("hancock.tar.gz", "r:gz")
>>> tar.extractall()
>>> tar.close()
```

We'll work on one more example by combining elements we've learned in this chapter as well as the elements in the *Vector data* section of *Chapter 2, Geospatial Data*. We'll read the bounding box coordinates from the `hancock.zip` file without ever saving it to disk. We'll use the power of Python's file-like object convention to pass around the data. Then, we'll use Python's `struct` module to read the bounding box as we did in Chapter 2. In this case, we read the unzipped `.shp` file into a variable and access the data using Python array slicing by specifying the starting and ending indexes of the data separated by a colon (`:`). We are able to use list slicing because Python allows you to treat strings as lists. In this example, we also use Python's `StringIO` module to temporarily store data in memory in a file-like object that implements all methods including the `seek()` method, which is absent from most Python networking modules, as shown in the following lines of code:

```
>>> import urllib.request
>>> import urllib.parse
>>> import urllib.error
>>> import zipfile
>>> import io
>>> import struct
>>> url =
    "https://github.com/GeospatialPython/Learn/raw/master/hancock.zip"
>>> cloudshape = urllib.request.urlopen(url)
>>> memoryshape = io.BytesIO(cloudshape.read())
>>> zipshape = zipfile.ZipFile(memoryshape)
```

```
>>> cloudshp = zipshape.read("hancock.shp")
# Access Python string as an array
>>> struct.unpack("<dddd", cloudshp[36:68])
(-89.6904544701547, 30.173943486533133, -89.32227546981174,
  30.6483914869749)
```

As you can see from the examples so far, Python's standard library packs a lot of punch. Most of the time, you don't have to download a third-party library just to access a file online.

Python markup and tag-based parsers

Tag-based data, particularly different XML dialects, have become a very popular way to distribute geospatial data. Formats that are both machine and human readable are generally easy to work with, though they sacrifice storage efficiency for usability. These formats can become unmanageable for very large datasets but work very well in most cases.

While most formats are some form of XML (such as KML or GML), there is a notable exception. The **well-known text (WKT)** format is fairly common but uses external markers and square brackets ([]) to surround data instead of tags in angled brackets around data like XML does.

Python has standard library support for XML as well as some excellent third-party libraries available. Proper XML formats all follow the same structure, so you can use a generic XML library to read it. Because XML is text-based, it is often easy to write it as a string instead of using an XML library. The vast majority of applications which output XML do so in this way. The primary advantage of using XML libraries for writing XML is that your output is usually validated. It is very easy to create an error while creating your own XML format. A single missing quotation mark can derail an XML parser and throw an error for somebody trying to read your data. When these errors happen, they virtually render your dataset useless. You will find that this problem is very common among XML-based geospatial data. What you'll discover is that some parsers are more forgiving with incorrect XML than others. Often, reliability is more important than speed or memory efficiency. The analysis available at `http://lxml.de/performance.html` provides benchmarks for memory and speed among the different Python XML parsers.

The minidom module

The Python `minidom` module is a very old and simple to use XML parser. It is a part of Python's built-in set of XML tools in the XML package. It can parse XML files or XML fed in as a string. The `minidom` module is best for small to medium-sized XML documents of less than about 20 megabytes before speed begins to decrease.

To demonstrate the `minidom` module, we'll use a sample KML file which is a part of Google's KML documentation that you can download. The data available at the following link represents time-stamped point locations transferred from a GPS device:

```
https://github.com/GeospatialPython/Learn/raw/master/time-stamp-
point.kml
```

First, we'll parse this data by reading it in from the file and creating a `minidom` parser object. The file contains a series of `<Placemark>` tags, which contain a point and a timestamp at which that point was collected. So, we'll get a list of all of the Placemarks in the file, and we can count them by checking the length of that list, as shown in the following lines of code:

```
>>> from xml.dom import minidom
>>> kml = minidom.parse("time-stamp-point.kml")
>>> Placemarks = kml.getElementsByTagName("Placemark")
>>> len(Placemarks)
361
```

As you can see, we retrieved all `Placemarks` which totaled `361`. Now, let's take a look at the first `Placemark` element in the list:

```
>>> Placemarks[0]
<DOM Element: Placemark at 0x2045a30>
```

Each `<Placemark>` tag is now a `DOM Element` data type. To really see what that element is, we call the `toxml()` method as you can see in the following lines of code:

```
>>> Placemarks[0].toxml()
u'<Placemark>\n <TimeStamp>\n \<when>2007-01-14T21:05:02Z</when>\n
</TimeStamp>\n <styleUrl>#paddle-a</styleUrl>\n <Point>\n
  <coordinates>-122.536226,37.86047,0</coordinates>\n
  </Point>\n </Placemark>'
```

The `toxml()` function outputs everything contained in the `Placemark` tag as a string object. If we want to print this information to a text file, we can call the `toprettyxml()` method, which would add additional indentation to make the XML more readable.

Now what if we want to grab just the coordinates from this Placemark? The coordinates are buried inside the `coordinates` tag, which is contained in the `Point` tag and nested inside the `Placemark` tag. Each element of a `minidom` object is called a node. Nested nodes are called children or child nodes. The child nodes include more than just tags. They can also include whitespace separating tags as well as the data inside tags. So, we can drill down to the `coordinates` tag using the tag name, but then we'll need to access the data node. All the `minidom` elements have a `childNodes` list as well as a `firstChild()` method to access the first node. We'll combine these methods to get to the `data` attribute of the first coordinate's data node, which we reference using index 0 in the list of `coordinates` tags:

```
>>> coordinates =
    Placemarks[0].getElementsByTagName("coordinates")
>>> point = coordinates[0].firstChild.data
>>> point
u'-122.536226,37.86047,0'
```

If you're new to Python, you'll notice that the text output in these examples is tagged with the letter u. This markup is how Python denotes Unicode strings which support internationalization to multiple languages with different character sets. Python 3.4.3 changes this convention slightly and leaves Unicode strings unmarked while marking utf-8 strings with a b.

We can go a little further and convert this `point` string into usable data by splitting the string and converting the resulting strings as Python float types, as shown here:

```
>>> x,y,z = point.split(",")
>>> x
u'-122.536226'
>>> y
u'37.86047'
>>> z
u'0'
>>> x = float(x)
>>> y = float(y)
>>> z = float(z)
>>> x,y,z
(-122.536226, 37.86047, 0.0)
```

Using a Python list comprehension, we can perform this operation in a single step, as you can see in the following lines of code:

```
>>> x,y,z = [float(c) for c in point.split(",")]
>>> x,y,z
(-122.536226, 37.86047, 0.0)
```

This example scratches the surface of what the `minidom` library can do. For a great tutorial on this library, have a look at the *9.3. Parsing XML* section of the excellent book *Dive Into Python*, by *Mark Pilgrim*, which is available in print or online at `http://www.diveintopython.net/xml_processing/parsing_xml.html`.

ElementTree

The `minidom` module is pure Python, easy to work with, and has been around since Python 2.0. However, Python 2.5 added a more efficient yet high-level XML parser to the standard library called `ElementTree`. `ElementTree` is interesting because it has been implemented in multiple versions. There is a pure Python version and a faster version written in C called `cElementTree`. You should use `cElementTree` wherever possible, but it's possible that you may be on a platform that doesn't include the C-based version. When you import `cElementTree`, you can test to see if it's available and fall back to the pure Python version if necessary:

```
try:
    import xml.etree.cElementTree as ET
except ImportError:
    import xml.etree.ElementTree as ET
```

One of the great features of `ElementTree` is its implementation of a subset of the XPath query language. XPath is short for XML Path and allows you to search an XML document using a path-style syntax. If you work with XML frequently, learning XPath is essential. You can learn more about XPath at the following link:

`http://www.w3schools.com/xsl/xpath_intro.asp`

One catch with this feature is if the document specifies a namespace, as most XML documents do, you must insert that namespace into queries. `ElementTree` does not automatically handle the namespace for you. Your options are to manually specify it or try to extract it using string parsing from the root element's tag name.

We'll repeat the `minidom` XML parsing example using `ElementTree`. First, we'll parse the document and then we'll manually define the KML namespace; later, we'll use an XPath expression and the `find()` method to find the first `Placemark` element. Finally, we'll find the coordinates and the child node and then grab the text containing the latitude and longitude. In both cases, we could have searched directly for the `coordinates` tag. But by grabbing the `Placemark` element, it gives us the option of grabbing the corresponding timestamp child element later, if we choose, as shown in the following lines of code:

```
>>> tree = ET.ElementTree(file="time-stamp-point.kml")
```

```
>>> ns = "{http://www.opengis.net/kml/2.2}"
>>> placemark = tree.find(".//%sPlacemark" % ns)
>>> coordinates =
    placemark.find("./{}Point/{}coordinates".format(ns, ns))
>>> coordinates.text
'-122.536226,37.86047,0'
```

In this example, notice that we used the Python string formatting syntax, which is based on the string formatting concept found in C. When we defined the XPath expression for the `placemark` variable, we used the `%s` placeholder to specify the insertion of a string. Then after the string, we use the `%` operator followed by a variable name to insert the `ns` namespace variable where the placeholder is. In the `coordinates` variable, we use the `ns` variable twice, so we specify a tuple containing `ns` twice after the string.

> String formatting is a simple yet extremely powerful and useful tool in Python, which is worth learning. You can find more information in Python's documentation online at the following link:
>
> https://docs.python.org/3.4/library/string.html:

Building XML

Most of the time, XML can be built by concatenating strings, as you can see in the following command:

```
xml = "<?xml version="1.0" encoding="utf-8"?>"
xml += "<kml xmlns="http://www.opengis.net/kml/2.2">"
xml += "   <Placemark>"
xml += "      <name>Office</name>"
xml += "      <description>Office Building</description>"
xml += "      <Point>"
xml += "         <coordinates>"
xml += "            -122.087461,37.422069"
xml += "         </coordinates>"
xml += "      </Point>"
xml += "   </Placemark>"
xml += "</kml>"
```

But, this method can be quite prone to typos, which creates invalid XML documents. A safer way is to use an XML library. Let's build this simple KML document using ElementTree. We'll define the root KML element and assign it a namespace. Then, we'll systematically append subelements to the root, and finally wrap the elements as an ElementTree object, declare the XML encoding, and write it out to a file called placemark.xml, as shown in the following lines of code:

```
>>> root = ET.Element("kml")
>>> root.attrib["xmlns"] = "http://www.opengis.net/kml/2.2"
>>> placemark = ET.SubElement(root, "Placemark")
>>> office = ET.SubElement(placemark, "name")
>>> office.text = "Office"
>>> point = ET.SubElement(placemark, "Point")
>>> coordinates = ET.SubElement(point, "coordinates")
>>> coordinates.text = "-122.087461,37.422069, 37.422069"
>>> tree = ET.ElementTree(root)
>>> tree.write("placemark.kml",
  xml_declaration=True,encoding='utf-8',method="xml")
```

The output is identical to the previous string building example, except that ElementTree does not indent the tags but rather writes it as one long string. The minidom module has a similar interface, which is documented in the book *Dive Into Python*, by *Mark Pilgrim*, referenced in the minidom example that we just saw.

XML parsers such as minidom and ElementTree work very well on perfectly formatted XML documents. Unfortunately, the vast majority of XML documents out there don't follow the rules and contain formatting errors or invalid characters. You'll find that you are often forced to work with this data and must resort to extraordinary string parsing techniques to get the small subset of data you actually need. But thanks to Python and BeautifulSoup, you can elegantly work with bad and even terrible, tag-based data.

BeautifulSoup is a module specifically designed to robustly handle broken XML. It is oriented towards HTML, which is notorious for incorrect formatting but works with other XML dialects too. BeautifulSoup is available on PyPI, so use either easy_install or pip to install it, as you can see in the following command:

```
easy_install beautifulsoup4
```

Or you can execute the following command:

```
pip install beautifulsoup4
```

Then, to use it, you simply import it:

```
>>> from bs4 import BeautifulSoup
```

To try it out, we'll use a **GPS Exchange Format** (**GPX**) tracking file from a smartphone application, which has a glitch and exports slightly broken data. You can download this sample file which is available at the following link:

```
https://raw.githubusercontent.com/GeospatialPython/Learn/master/
broken_data.gpx
```

This 2,347 line data file is in pristine condition except that it is missing a closing `</trkseg>` tag, which should be located at the very end of the file just before the closing `</trk>` tag. This error was caused by a data export function in the source program. This defect is most likely a result of the original developer manually generating the GPX XML on export and forgetting the line of code that adds this closing tag. Watch what happens if we try to parse this file with `minidom`:

```
>>> gpx = minidom.parse("broken_data.gpx")
Traceback (most recent call last):
  File "<stdin>", line 1, in <module>
  File "C:\Python34\lib\xml\dom\minidom.py", line 1914, in parse
    return expatbuilder.parse(file)
  File "C:\Python34\lib\xml\dom\expatbuilder.py", line 924, in
    parse
    result = builder.parseFile(fp)
  File "C:\Python34\lib\xml\dom\expatbuilder.py", line 207, in
    parseFile
    parser.Parse(buffer, 0)
xml.parsers.expat.ExpatError: mismatched tag: line 2346, column 2
```

As you can see from the last line in the error message, the underlying XML parser in `minidom` knows exactly what the problem is—a mismatched tag right at the end of the file. But, it refused to do anything more than report the error. You must have perfectly formed XML or none at all to avoid this.

Now, let's try the more sophisticated and efficient `ElementTree` module with the same data:

```
>>> ET.ElementTree(file="broken_data.gpx")
Traceback (most recent call last):
  File "<stdin>", line 1, in <module>
  File "C:\Python34\lib\xml\etree\ElementTree.py", line 611, in
    __init__
    self.parse(file)
  File "C:\Python34\lib\xml\etree\ElementTree.py", line 653, in
    parse
    parser.feed(data)
  File "C:\Python34\lib\xml\etree\ElementTree.py", line 1624, in
    feed
```

```
        self._raiseerror(v)
      File "C:\Python34\lib\xml\etree\ElementTree.py", line 1488, in
      _raiseerror
        raise err
    xml.etree.ElementTree.ParseError: mismatched tag: line 2346,
      column 2
```

As you can see, different parsers face the same problem. Poorly formed XML is an all too common reality in geospatial analysis, and every XML parser assumes that all the XML in the world is perfect, except for one. Enter `BeautifulSoup`. This library shreds bad XML into usable data without a second thought. And it can handle far worse defects than missing tags. It will work despite missing punctuation or other syntax and will give you the best data it can. It was originally developed for parsing HTML, which is notoriously horrible for being poorly formed, but it works fairly well with XML as well, as shown here:

```
>>> from bs4 import BeautifulSoup
>>> gpx = open("broken_data.gpx")
>>> soup = BeautifulSoup(gpx.read(), features="xml")
>>>
```

No complaints from `BeautifulSoup`! Just to make sure the data is actually usable, let's try and access some of the data. One of the fantastic features of `BeautifulSoup` is that it turns tags into attributes of the parse tree. If there are multiple tags with the same name, it grabs the first one. Our sample data file has hundreds of `<trkpt>` tags. Let's access the first one:

```
>>> soup.trkpt
<trkpt lat="30.307267000" lon="-89.332444000">
  <ele>10.7</ele><time>2013-05-16T04:39:46Z</time></trkpt>
```

We're now certain that the data has been parsed correctly and we can access it. If we want to access all of the `<trkpt>` tags, we can use the `findAll()` method to grab them and then use the built-in Python `len()` function to count them, as shown here:

```
>>> tracks = soup.findAll("trkpt")
>>> len(tracks)
2321
```

If we write the parsed data back out to a file, `BeautifulSoup` outputs the corrected version. We'll save the fixed data as a new GPX file using `BeautifulSoup` module's `prettify()` method to format the XML with nice indentation, as you can see in the following lines of code:

```
>>> fixed = open("fixed_data.gpx", "w")
>>> fixed.write(soup.prettify())
>>> fixed.close()
```

`BeautifulSoup` is a very rich library with many more features. To explore it further, visit the `BeautifulSoup` documentation online at `http://www.crummy.com/ software/BeautifulSoup/bs4/documentation.html`.

> While `minidom`, `ElementTree`, and `cElementTree` come with the Python standard library, there is an even more powerful and popular XML library for Python called `lxml`. The `lxml` module provides a Pythonic interface to the `libxml2` and `libxslt` C libraries using the `ElementTree` API. An even better fact is that `lxml` also works with `BeautifulSoup` to parse bad tag-based data. On some installations, `BeautifulSoup4` may require `lxml`. The `lxml` module is available via PyPI but requires some additional steps for the C libraries. More information is available on the `lxml` homepage at the following link:
>
> `http://lxml.de/`

Well-known text (WKT)

The WKT format has been around for years and is a simple text-based format for representing geometries and spatial reference systems. It is primarily used as a data exchange format by systems, which implement the OGC Simple Features for SQL specification. Take a look at the following sample WKT representation of a polygon:

```
POLYGON((0 0,4 0,4 4,0 4,0 0),(1 1, 2 1, 2 2, 1 2,1 1))
```

Currently, the best way to read and write WKT is the **Shapely** library. Shapely provides a very Python-oriented or Pythonic interface to the **Geometry Engine – Open Source (GEOS)** library described in *Chapter 3, The Geospatial Technology Landscape*.

You can install Shapely using either `easy_install` or `pip`. You can also use the wheel from the site mentioned in the previous section. Shapely has a WKT module which can load and export this data. Let's use Shapely to load the previous polygon sample and then verify that it has been loaded as a polygon object by calculating its area:

```
>>> import shapely.wkt
>>> wktPoly = "POLYGON((0 0,4 0,4 4,0 4,0 0),(1 1, 2 1, 2 2, 1 2,
  1 1))"
>>> poly = shapely.wkt.loads(wktPoly)
>>> poly.area
15.0
```

We can convert any Shapely geometry back to a WKT by simply calling its `wkt` attribute, as shown here:

```
>>> poly.wkt
'POLYGON ((0.0 0.0, 4.0 0.0, 4.0 4.0, 0.0 4.0, 0.0 0.0), (1.0 1.0,
  2.0 1.0, 2.0 2.0, 1.0 2.0, 1.0 1.0))'
```

Shapely can also handle the WKT binary counterpart called **Well-known binary (WKB)** used to store WKT strings as binary objects in databases. Shapely loads WKB using its `wkb` module in the same way as the `wkt` module, and it can convert geometries by calling that object's `wkb` attribute.

Shapely is the most Pythonic way to work with WKT data, but you can also use the Python bindings to the OGR library, which we installed earlier in this chapter.

For this example, we'll use a shapefile with one simple polygon, which can be downloaded as a ZIP file available at the following link: `https://github.com/GeospatialPython/Learn/raw/master/polygon.zip`

In the following example, we'll open the `polygon.shp` file from the `shapefile` dataset, call the required `GetLayer()` method, get the first (and only) feature, and then export it to WKT:

```
>>> from osgeo import ogr
>>> shape = ogr.Open("polygon.shp")
>>> layer = shape.GetLayer()
>>> feature = layer.GetNextFeature()
>>> geom = feature.GetGeometryRef()
>>> wkt = geom.ExportToWkt()
>>> wkt
' POLYGON ((-99.904679362176353 51.698147686745074,
  -75.010398603076666 46.56036851832075,-75.010398603076666
  46.56036851832075,-75.010398603076666 46.56036851832075,
  -76.975736557742451 23.246272688996914,-76.975736557742451
  23.246272688996914,-76.975736557742451 23.246272688996914,
  -114.31715769639194 26.220870210283724,-114.31715769639194
  26.220870210283724,-99.904679362176353 51.698147686745074))'
```

Note that with OGR, you would have to read access each feature and export it individually as the `ExporttoWkt()` method is at the feature level. We can now turn around and read a WKT string using the `wkt` variable containing the export. We'll import it back into `ogr` and get the bounding box, also known as an envelope, of the polygon, as you can see here:

```
>>> poly = ogr.CreateGeometryFromWkt(wkt)
>>> poly.GetEnvelope()
(-114.31715769639194, -75.01039860307667, 23.246272688996914,
  51.698147686745074)
```

Shapely and OGR are used for reading and writing valid WKT strings. Of course, just like XML, which is also text, you could manipulate small amounts of WKT as strings in a pinch.

Python JSON libraries

JavaScript Object Notation (JSON) is rapidly becoming the number one data exchange format across a lot of fields. The lightweight syntax and the similarity to existing data structures in both the JavaScript from which Python borrows some data structures makes it a perfect match for Python.

The following **GeoJSON** sample document contains a single point:

```
{
    "type": "Feature",
    "id": "OpenLayers.Feature.Vector_314",
    "properties": {},
    "geometry": {
        "type": "Point",
        "coordinates": [
            97.03125,
            39.7265625
        ]
    },
    "crs": {
        "type": "name",
        "properties": {
            "name": "urn:ogc:def:crs:OGC:1.3:CRS84"
        }
    }
}
```

This sample is just a simple point with new attributes, which would be stored in the `properties` data structure of the geometry. First, we'll compact the sample document into a single string to make it easier to handle:

```
>>> jsdata = """{ "type": "Feature", "id":
  "OpenLayers.Feature.Vector_314", "pro
perties": {}, "geometry": { "type": "Point", "coordinates": [
97.03125, 39.72656
25 ] }, "crs": { "type": "name", "properties": { "name":
  "urn:ogc:def:crs:OGC:1.
3:CRS84" } } }"""
```

The json module

GeoJSON looks very similar to a nested set of Python's dictionaries and lists. Just for fun, let's just try and use Python's `eval()` function to parse it as Python code:

```
>>> point = eval(jsdata)
>>> point["geometry"]
{'type': 'Point', 'coordinates': [97.03125, 39.7265625]}
```

Wow! That just worked! We turned that random GeoJSON string into native Python data in one easy step. Keep in mind that the JSON data format is based on JavaScript syntax which happens to be similar to Python. Also, as you get deeper into GeoJSON data and work with larger data, you'll find that JSON allows characters that Python does not. Using Python's `eval()` function is considered very insecure as well. But as far as keeping things simple is concerned, note that it doesn't get any simpler than that!

Thanks to Python's drive towards simplicity, the more advanced method doesn't get much more complicated. Let's use Python's `json` module, which is part of the standard library, to turn the same string into Python the right way:

```
>>> import json
>>> json.loads(jsdata)
{u'geometry': {u'type': u'Point', u'coordinates': [97.03125,
    39.7265625]}, u'crs
': {u'type': u'name', u'properties': {u'name':
    u'urn:ogc:def:crs:OGC:1.3:CRS84'}
}, u'type': u'Feature', u'id': u'OpenLayers.Feature.Vector_314',
    u'properties':
{}}
```

As a side note, in the previous example the CRS84 property is a synonym for the common WGS84 coordinate system. The `json` module adds some nice features such as safer parsing and conversion of strings to Unicode. We can export Python data structures to JSON in almost the same way:

```
>>> pydata = json.loads(jsdata)
>>> json.dumps(pydata)
'{"geometry": {"type": "Point", "coordinates": [97.03125,
    39.7265625]}, "crs": {
"type": "name", "properties": {"name":
    "urn:ogc:def:crs:OGC:1.3:CRS84"}}, "type"
: "Feature", "id": "OpenLayers.Feature.Vector_314", "properties":
    {}}'
```

The geojson module

We could happily go on reading and writing GeoJSON data using the `json` module forever, but there's an even better way. The `geojson` module available on PyPI offers some distinct advantages. For starters, it knows the requirements of the GeoJSON specification, which can save a lot of typing. Let's create a simple point using this module and export it to GeoJSON:

```
>>> import geojson
>>> p = geojson.Point([-92, 37])
>>> geojs = geojson.dumps(p)
>>> geojs
'{"type": "Point", "coordinates": [-92, 37]}'
```

Notice that the `geojson` module has an interface for different data types and saves us from setting the `type` and `coordinates` attributes manually. Now imagine if you had a geographic object with hundreds of features. You could programmatically build this data structure instead of building a very large string. The `geojson` module is also the reference implementation for the Python `__geo_interface__` convention. This interface allows cooperating programs to exchange data seamlessly and in a Pythonic way without the programmer explicitly exporting and importing GeoJSON strings. So, if we wanted to feed the point we created with the `geojson` module to the Shapely module, we could perform the following command, which reads the `geojson` module's `point` object straight into Shapely after which we'll export it as WKT:

```
>>> from shapely.geometry import asShape
>>> point = asShape(p)
>>> point.wkt
'POINT (-92.0000000000000000 37.0000000000000000)'
```

More and more geospatial Python libraries are implementing both the `geojson` and `__geo_interface__` functionality including PyShp, Fiona, Karta, and ArcGIS. Third-party implementations exist for QGIS.

OGR

We touched on OGR as a way to handle WKT strings, but its real power is as a universal vector library. This book strives for pure Python solutions, but no single library even comes close to the variety of formats that OGR can process.

Let's read a sample `point` shapefile using the OGR Python API. The sample shapefile can be downloaded as a ZIP file here: `https://github.com/GeospatialPython/Learn/raw/master/point.zip`

This `point` shapefile has five points with single digit, positive coordinates. The attributes list the order in which the points were created, making it useful for testing. This simple example will read in the `point` shapefile and loop through each feature; then it'll print the x and y value of each point plus the value of the first attribute field:

```
>>> from osgeo import ogr
>>> shp = ogr.Open("point.shp")
>>> layer = shp.GetLayer()
>>> for feature in layer:
...     geometry = feature.GetGeometryRef()
...     print(geometry.GetX(), geometry.GetY(),
...         feature.GetField("FIRST_FLD"))
...
1.0 1.0 First
3.0 1.0 Second
4.0 3.0 Third
2.0 2.0 Fourth
0.0 0.0 Appended
```

This example is simple, but OGR can become quite verbose as your script becomes more complex.

PyShp

`PyShp` is a simple, pure Python library that reads and writes shapefiles. It doesn't perform any geometry operations and only uses Python's standard library. It's contained in a single file that's easy to move around, squeeze onto small embedded platforms, and modify. It is also compatible with Python 3. It also implements `__geo_interface__`. The `PyShp` module is available on PyPI.

Let's repeat the previous OGR example with PyShp:

```
>>> import shapefile
>>> shp = shapefile.Reader("point.shp")
>>> for feature in shp.shapeRecords():
...     point = feature.shape.points[0]
...     rec = feature.record[0]
...     print(point[0], point[1], rec)
...
1.0 1.0 First
3.0 1.0 Second
4.0 3.0 Third
2.0 2.0 Fourth
0.0 0.0 Appended
```

dbfpy

Both OGR and `PyShp` read and write the `dbf` files because they are part of the shapefile specification. The `dbf` files contain the attributes and fields for the shapefiles. However, both libraries have very basic `dbf` support. Occasionally, you will need to do some heavy duty `dbf` work. The `dbfpy3` module is a pure Python module dedicated to working with `dbf` files. It is currently hosted on `github.com` only. You can force `easy_install` to find the download by specifying the download file:

```
easy_install -f
    https://github.com/GeospatialPython/dbfpy3/archive/master.zip
```

If you are using `pip` to install packages, use the following command:

```
pip install
    https://github.com/GeospatialPython/dbfpy3/archive/master.zip
```

The following shapefile has over 600 dbf records representing U.S. Census Bureau tracts which make it a good sample for trying out `dbfpy`:

https://github.com/GeospatialPython/Learn/raw/master/GIS_CensusTract.zip

Let's open up the `dbf` file of this shapefile and look at the first record:

```
>>> from dbfpy3 import dbf
>>> db = dbf.Dbf("GIS_CensusTract_poly.dbf")
>>> db[0]
  GEODB_OID: 4029 (<type 'int'>)
  OBJECTID: 4029 (<type 'int'>)
  PERMANE0: 61be9239-8f3b-4876-8c4c-0908078bc597 (<type 'str'>)
  SOURCE_1: NA (<type 'str'>)
  SOURCE_2: 20006 (<type 'str'>)
  SOURCE_3: Census Tracts (<type 'str'>)
  SOURCE_4: Census Bureau (<type 'str'>)
  DATA_SE5: 5 (<type 'str'>)
  DISTRIB6: E4 (<type 'str'>)
  LOADDATE: 2007-03-13 (<type 'datetime.date'>)
  QUALITY: 2 (<type 'str'>)
  SCALE: 1 (<type 'str'>)
  FCODE: 1734 (<type 'str'>)
  STCO_FI7: 22071 (<type 'str'>)
  STATE_NAME: 22 (<type 'str'>)
  COUNTY_8: 71 (<type 'str'>)
  CENSUST9: 22071001734 (<type 'str'>)
  POPULAT10: 1760 (<type 'int'>)
```

```
    AREASQKM: 264.52661934 (<type 'float'>)
    GNIS_ID: NA (<type 'str'>)
    POPULAT11: 1665 (<type 'int'>)
    DB2GSE_12: 264526619.341 (<type 'float'>)
    DB2GSE_13: 87406.406192 (<type 'float'>)
```

The module very quickly and easily gives us both the column names and data values. Now let's increment the population field contained in POPULAT10 by 1:

```
>>> rec = db[0]
>>> field = rec["POPULAT10"]
>>> rec["POPULAT10"] = field + 1
>>> rec.store()
>>> del rec
>>> db[0]["POPULAT10"]
1761
```

Keep in mind that both OGR and PyShp can do this same procedure, but dbfp3y makes it a little easier if you are making a lot of changes to the dbf files only.

Shapely

Shapely was mentioned in the *Well-known text (WKT)* section for import and export ability. But, its true purpose is a generic geometry library. Shapely is a high-level, Pythonic interface to the GEOS library for geometric operations. In fact, Shapely intentionally avoids reading or writing files. It relies completely on data import and export from other modules and maintains focus on geometry manipulation.

Let's do a quick Shapely demonstration in which we'll define a single WKT polygon and then import it into Shapely. Then we'll measure the area. Our computational geometry will consist of buffering that polygon by a measure of five arbitrary units, which will return a new, bigger polygon for which we'll measure the area:

```
>>> from shapely import wkt, geometry
>>> wktPoly = "POLYGON((0 0,4 0,4 4,0 4,0 0))"
>>> poly = wkt.loads(wktPoly)
>>> poly.area
16.0
>>> buf = poly.buffer(5.0)
>>> buf.area
174.41371226364848
```

We can then perform a difference on the area of the buffer and the original polygon area, as shown here:

```
>>> buf.difference(poly).area
158.413712264
```

If you can't have pure Python, a Pythonic API as clean as Shapely that packs such a punch is certainly the next best thing.

Fiona

The Fiona library provides a simple Python API around the OGR library for data access and nothing more. This approach makes it easy to use and is less verbose than OGR while using Python. Fiona outputs GeoJSON by default. You can find a wheel file for Fiona at http://www.lfd.uci.edu/~gohlke/pythonlibs/#fiona.

As an example, we'll use the `GIS_CensusTract_poly.shp` file from the `dbfpy` example seen earlier in this chapter.

First we'll import `fiona` and Python's `pprint` module to format the output. Then, we'll open the shapefile and check its driver type:

```
>>> import fiona
>>> import pprint
>>> f = fiona.open("GIS_CensusTract_poly.shp")
>>> f.driver
ESRI Shapefile
```

Next, we'll check its coordinate reference system and get the data bounding box, as shown here:

```
>>> f.crs
{'init': 'epsg:4269'}
>>> f.bounds
(-89.8744162216216, 30.161122135135138, -89.1383837783784,
  30.661213864864862)
```

Now, we'll view the data schema as `geojson` and format it using the `pprint` module, as you can see in the following lines of code:

```
>>> pprint.pprint(f.schema)
{'geometry': 'Polygon',
 'properties': {'GEODB_OID': 'float:11',
                'OBJECTID': 'float:11',
                'PERMANE0': 'str:40',
                'SOURCE_1': 'str:40',
```

```
                         'SOURCE_2': 'str:40',
                         'SOURCE_3': 'str:100',
                         'SOURCE_4': 'str:130',
                         'DATA_SE5': 'str:46',
                         'DISTRIB6': 'str:188',
                         'LOADDATE': 'date',
                         'QUALITY': 'str:35',
                         'SCALE': 'str:52',
                         'FCODE': 'str:38',
                         'STCO_FI7': 'str:5',
                         'STATE_NAME': 'str:140',
                         'COUNTY_8': 'str:60',
                         'CENSUST9': 'str:20',
                         'POPULAT10': 'float:11',
                         'AREASQKM': 'float:31.15',
                         'GNIS_ID': 'str:10',
                         'POPULAT11': 'float:11',
                         'DB2GSE_12': 'float:31.15',
                         'DB2GSE_13': 'float:31.15'}}
```

Next let's get a count of the number of features:

```
>>> len(list(f))
45
```

Finally, we'll print one of the records as formatted GeoJSON, as shown here:

```
pprint.pprint(f[1])
{'geometry': {'coordinates': [[[(-89.86412366375093,
   30.661213864864862), (-89.86418691770497, 30.660764012731285),
(-89.86443391770518, 30.659652012730202),
...
'type': 'MultiPolygon'},
'id': '1',
'properties': {'GEODB_OID': 4360.0,
               'OBJECTID': 4360.0,
               'PERMANE0': '9a914eef-9249-44cf-a05f-af4b48876c59',
               'SOURCE_1': 'NA',
               'SOURCE_2': '20006',
...
               'DB2GSE_12': 351242560.967882,
               'DB2GSE_13': 101775.283967268},
  'type': 'Feature'}
```

GDAL

GDAL is the dominant geospatial library for raster data. Its raster capability is so significant that it is a part of virtually every geospatial toolkit in any language, and Python is no exception to this. To see the basics of how GDAL works in Python, download the following sample raster satellite image as a ZIP file and unzip it: `https://github.com/GeospatialPython/Learn/raw/master/SatImage.zip`. Let's open this image and see how many bands it has and how many pixels are present along each axis:

```
>>> from osgeo import gdal
>>> raster = gdal.Open("SatImage.tif")
>>> raster.RasterCount
3
>>> raster.RasterXSize
2592
>>> raster.RasterYSize
2693
```

So, we see that the following image has three bands, 2,592 columns of pixels, and 2,693 rows of pixels by viewing it in **OpenEV**:

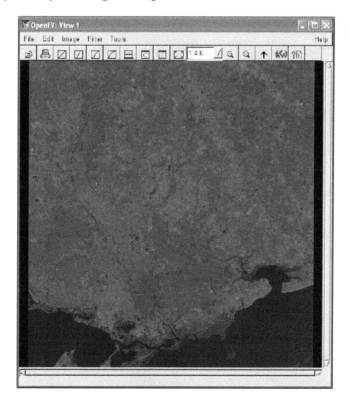

GDAL is an extremely fast geospatial raster reader and writer within Python. It can also reproject images quite well in addition to a few other tricks. However, the true value of GDAL comes from its interaction with the next Python module, which we'll examine now.

NumPy

NumPy is an extremely fast, multidimensional Python array processor designed specifically for Python and scientific computing but is written in C. It is available via PyPI or as a wheel file (available at http://www.lfd.uci.edu/~gohlke/pythonlibs/#numpy) and installs easily. In addition to its amazing speed, the magic of NumPy includes its interaction with other libraries. NumPy can exchange data with GDAL, Shapely, the **Python Imaging Library** (**PIL**), and many other scientific computing Python libraries in other fields.

As a quick example of NumPy's ability, we'll combine it with GDAL to read in our sample satellite image and then create a histogram of it. The interface between GDAL and NumPy is a GDAL module called `gdal_array`, which has NumPy as a dependency. Numeric is the legacy name of the NumPy module. The `gdal_array` module imports NumPy.

In the following example, we'll use `gdal_array`, which imports NumPy, to read the image in as an array, grab the first band, and save it back out as a JPEG image:

```
>>> from osgeo import gdal_array
>>> srcArray = gdal_array.LoadFile("SatImage.tif")
>>> band1 = srcArray[0]
>>> gdal_array.SaveArray(band1, "band1.jpg", format="JPEG")
```

This operation gives us the following grayscale image in OpenEV:

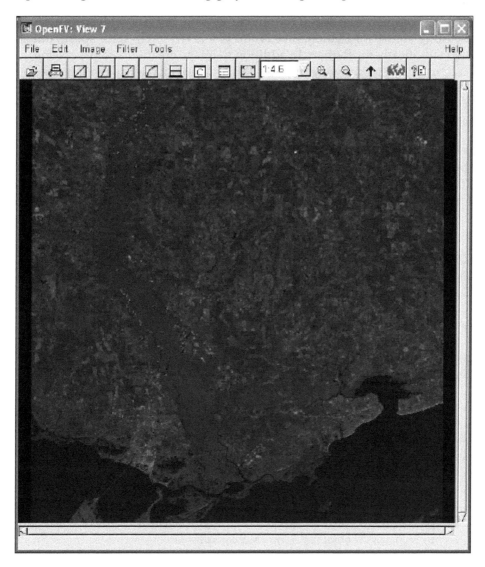

PIL

The PIL was originally developed for remote sensing, but has evolved as a general image editing library for Python. Like NumPy, it is written in C for speed, but is designed specifically for Python. In addition to image creation and processing, it also has a useful raster drawing module. PIL is also available via PyPI; however, in Python 3, you may want to use the `Pillow` module which is an upgraded version of PIL. As you'll see in the example below, we use a Python `try` statement to import PIL using two possible variations depending on how you installed it.

In this example, we'll combine PyShp and PIL to rasterize the `hancock` shapefile from previous examples and save it as an image. We'll use a *world to pixel* coordinate transformation similar to our SimpleGIS from *Chapter 1, Learning Geospatial Analysis with Python*. We'll create an image to use as a canvas in PIL, and then we'll use the PIL `ImageDraw` module to render the polygon. Finally, we'll save it as a PNG image, as you can see in the following lines of code:

```
>>> try:
>>>     import Image
>>>     import ImageDraw
>>> except:
>>>     from PIL import Image
>>>     from PIL import ImageDraw
>>> import shapefile
>>> r = shapefile.Reader("hancock.shp")
>>> xdist = r.bbox[2] - r.bbox[0]
>>> ydist = r.bbox[3] - r.bbox[1]
>>> iwidth = 400
>>> iheight = 600
>>> xratio = iwidth/xdist
>>> yratio = iheight/ydist
>>> pixels = []
>>> for x,y in r.shapes()[0].points:
...     px = int(iwidth - ((r.bbox[2] - x) * xratio))
...     py = int((r.bbox[3] - y) * yratio)
...     pixels.append((px,py))
...
>>> img = Image.new("RGB", (iwidth, iheight), "white")
>>> draw = ImageDraw.Draw(img)
>>> draw.polygon(pixels, outline="rgb(203, 196, 190)",
  fill="rgb(198, 204, 189)")
>>> img.save("hancock.png")
```

This example creates the following image:

PNGCanvas

Sometimes, you may find that PIL is overkill for your purposes, or you are not allowed to install PIL because you do not have administrative rights to the machine that you're using to install Python modules created and compiled in C. In those cases, you can usually get away with the lightweight pure Python PNGCanvas module. You can install it using easy_install or pip.

Using this module, we can repeat the raster shapefile example we performed using PIL but in pure Python, as you can see here:

```
>>> import shapefile
>>> import pngcanvas
>>> r = shapefile.Reader("hancock.shp")
>>> xdist = r.bbox[2] - r.bbox[0]
>>> ydist = r.bbox[3] - r.bbox[1]
>>> iwidth = 400
>>> iheight = 600
>>> xratio = iwidth/xdist
>>> yratio = iheight/ydist
>>> pixels = []
>>> for x,y in r.shapes()[0].points:
...     px = int(iwidth - ((r.bbox[2] - x) * xratio))
...     py = int((r.bbox[3] - y) * yratio)
...     pixels.append([px,py])
...
>>> c = pngcanvas.PNGCanvas(iwidth,iheight)
>>> c.polyline(pixels)
>>> f = open("hancock_pngcvs.png", "wb")
>>> f.write(c.dump())
>>> f.close()
```

This example gives us a simple outline as `PNGCanvas` does not have a built-in fill method:

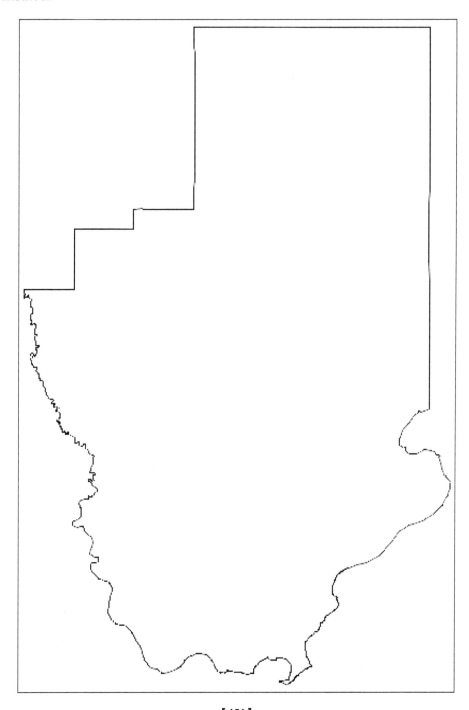

GeoPandas

Pandas is a high-performance Python data analysis library, which can handle large datasets that are tabular (similar to a database), ordered/unordered, labeled matrices, or unlabeled statistical data. GeoPandas is simply a geospatial extension to Pandas that builds upon Shapely, Fiona, PyProj, matplotlib, and Descartes, all of which must be installed. It enables you to easily perform operations in Python, which would otherwise require a spatial database such as PostGIS. You can download a wheel file for GeoPandas from `http://www.lfd.uci.edu/~gohlke/pythonlibs/#panda`.

The following script opens a shapefile and dumps it to GeoJSON; it then creates a map with `matplotlib`:

```
>>> import geopandas
>>> import matplotlib.pyplot as plt
>>> gdf = geopandas.GeoDataFrame
>>> census = gdf.from_file("GIS_CensusTract_poly.shp")
>>> census.plot()
>>> plt.show()
```

The following image is the resulting map plot of the previous commands:

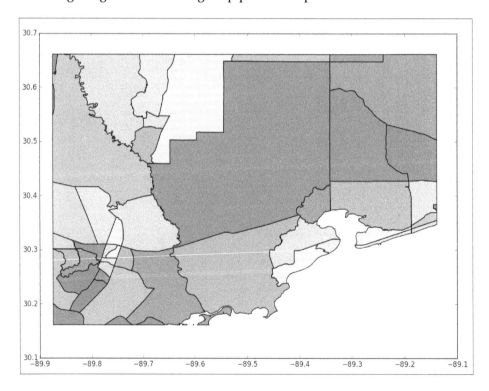

PyMySQL

The popular MySQL (available at `http://dev.mysql.com/downloads`) database is gradually evolving spatial functions. It has support for OGC geometries and a few spatial functions. It also has a pure Python API available in the **PyMySQL** library. The limited spatial functions use planar geometry and bounding rectangles as opposed to spherical geometry and shapes. The latest development release of MySQL contains some additional functions that improve this capability.

In the following example, we'll create a database in MySQL called `spatial_db`. Then, we'll add a table called `PLACES` with a `geometry` column. Next, we'll add two cities as point locations. And finally, we'll calculate the distance using MySQL's `ST_Distance` function and then convert the result from degrees to miles:

```
>>> import pymysql

>>> conn = pymysql.connect(host='localhost', port=3306,
  user='root', passwd='', db='mysql')
>>> cur = conn.cursor()
>>> cur.execute("DROP DATABASE IF EXISTS spatial_db")
>>> cur.execute("CREATE DATABASE spatial_db")
>>> cur.close()
>>> conn.close()
>>> conn = pymysql.connect(host='localhost', port=3306,
  user='root', passwd='', db='spatial_db')
>>> cur = conn.cursor()
>>> cur.execute("CREATE TABLE PLACES (id int NOT NULL
  AUTO_INCREMENT PRIMARY KEY, Name varchar(50) NOT NULL, location
  Geometry NOT NULL)")
>>> cur.execute("INSERT INTO PLACES (name, location) VALUES ('NEW
  ORLEANS', GeomFromText('POINT(30.03 90.03)'))")
>>> cur.execute("INSERT INTO PLACES (name, location) VALUES
  ('MEMPHIS', GeomFromText('POINT(35.05 90.00)'))")
>>> conn.commit()
>>> cur.execute("SELECT AsText(location) FROM PLACES")
>>> p1, p2 = [p[0] for p in cur.fetchall()]
>>> cur.execute("SET @p1 = ST_GeomFromText('{}')".format(p1))
>>> cur.execute("SET @p2 = ST_GeomFromText('{}')".format(p2))
>>> cur.execute("SELECT ST_Distance(@p1, @p2)")
>>> d = float(cur.fetchone()[0])
>>> print("{:.2f} miles from New Orleans to Memphis".format(d *
  70))
>>> cur.close()
>>> conn.close()
351.41 miles from New Orleans to Memphis
```

There are other spatial database options including PostGIS and SpatiaLite; however, Python 3 support for these spatial engines is developmental at best. You can access PostGIS and MySQL through the OGR library; however, MySQL support is limited.

PyFPDF

The pure Python **PyFPDF** library is a lightweight way to create PDFs including maps. Because the PDF format is a widely used standard, PDFs are commonly used to distribute maps. You can install it via PyPI as `fpdf`. The official name of the software is PyFPDF because it is a part of the PHP language module called `fpdf`. This module uses a concept called a cell to layout items at specific locations on a page. As a quick example, we'll import our `hancock.png` image created from the PIL example into a PDF called `map.pdf` to create a simple PDF map. The map will have the header text at the top that says **Hancock County Boundary**, followed by the map image:

```
>>> import fpdf
>>> # PDF constructor:
>>> # Portrait, millimeter units, A4 page size
>>> pdf=fpdf.FPDF("P", "mm", "A4")
>>> # create a new page
>>> pdf.add_page()
>>> # Set font: arial, bold, size 20
>>> pdf.set_font('Arial','B',20)
>>> # Layout cell: 160 x 25mm, title, no border, centered
>>> pdf.cell(160,25,'Hancock County Boundary', \
>>> border=0, align="C")
>>> # Write the image specifying the size
>>> pdf.image("hancock.png",25,50,110,160)
>>> # Save the file: filename, F = to file System
>>> pdf.output('map.pdf','F')
```

If you open the PDF file named `map.pdf` in Adobe Acrobat Reader or another PDF reader such as **SumatraPDF**, you'll see that the image is now centered on an A4 page. Geospatial products are often included as part of larger reports, and the PyFPDF module simplifies automatically generating reports as PDFs.

Spectral Python

Spectral Python (SPy) is a very advanced Python package for remote sensing. It goes far beyond what you would typically do with GDAL and NumPy and focuses on hyperspectral processing for images, which may have hundreds of bands. The basic package installs easily via PyPI, but SPy can provide a fully windowed processing environment if you install some additional dependencies. The remote sensing we'll perform in the rest of this book won't require SPy, but it is worth mentioning here because it is a well-maintained, powerful package that is competitive with many commercial software products in this domain. You can find out more about SPy at the homepage available at `http://spectralpython.sourceforge.net/index.html`.

Summary

In this chapter, we surveyed the Python-specific tools for geospatial analysis. Many of these tools included bindings to the libraries discussed in *Chapter 3, The Geospatial Technology Landscape* for best-of-breed solutions for specific operations like GDAL's raster access functions. We also included pure Python libraries as much as possible and will continue to include pure Python algorithms as we work through the upcoming chapters. In the next chapter, we'll begin applying these tools for GIS analysis.

5
Python and Geographic Information Systems

This chapter will focus on applying Python to functions typically performed by a **geographic information system** (**GIS**) such as QGIS or ArcGIS. We will continue to use as few external dependencies as possible outside Python itself, so you have tools which are as reusable as possible in different environments. In this book, we separate GIS analysis and remote sensing from programming perspective, which means that in this chapter, we'll focus on mostly vector data.

As with other chapters in this book, the items presented here are core functions that serve as building blocks which can be recombined to solve challenges you will encounter beyond this book. This chapter includes the following topics:

- Measuring distance
- Converting coordinates
- Reprojecting vector data
- Editing shapefiles
- Selecting data from within larger datasets
- Creating thematic maps
- Conversion of non-GIS data types
- Geocoding

This chapter contains many code samples. In addition to the text, code comments are included as guides within the samples.

Measuring distance

The essence of geospatial analysis is discovering the relationships of objects on the Earth. Items which are closer together tend to have a stronger relationship than those which are farther apart. This concept is known as **Tobler's First Law of Geography**. Therefore, measuring distance is a critical function of geospatial analysis.

As you have learned, every map is a model of the Earth, and they are all wrong to some degree. For this reason, measuring accurate distance between two points on the Earth while sitting in front of a computer is impossible. Even professional land surveyors who go out in the field with both traditional sighting equipment and very precise GPS equipment fail to account for every anomaly on the Earth's surface between point A and point B. So, in order to measure distance, we must look at what we are measuring, how much we are measuring, and how much accuracy we need.

There are three models of the Earth we can use to calculate distance:

- Flat plane
- Spherical
- Ellipsoid

In the flat plane model, standard Euclidean geometry is used. The Earth is considered a flat plane with no curvature as shown in the following figure:

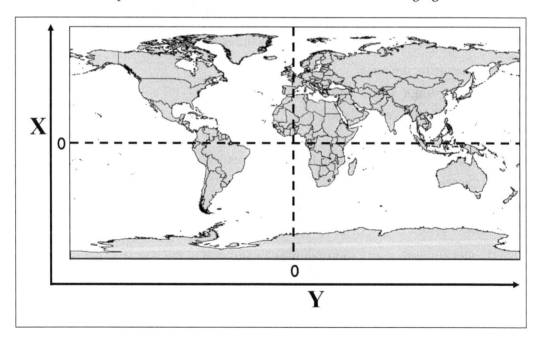

This model makes math quite simple because you work with straight lines. The most common format for geospatial coordinates is decimal degrees. However, decimal degree coordinates are reference measurements on a sphere taken as angles between the longitude and the prime meridian, and the latitude and equator. Furthermore, the lines of longitude converge toward zero at the poles. The circumference of each line of latitude becomes smaller toward the poles as well. These facts mean decimal degrees are not a valid coordinate system for Euclidean geometry, which uses infinite planes.

Map projections attempt to simplify the issues of dealing with a three-dimensional ellipsoid in a two-dimensional plane, which could be either a paper or a computer screen. As discussed in *Chapter 1, Learning Geospatial Analysis with Python*, map projections flatten a round model of the Earth to a plane and introduce distortion in exchange for the convenience of a map. Once this projection is in place and decimal degrees are traded for a **Cartesian** coordinate system with x and y coordinates, we can use the simplest forms of Euclidean geometry – the Pythagorean theorem.

At a large enough scale, a sphere or ellipsoid like the Earth, appears more like a plane than a sphere. In fact, for centuries, everyone thought that the Earth was flat! If the difference in degrees of longitude is small enough, you can often get away with using Euclidean geometry and then converting the measurements to meters, kilometers, or miles. This method is generally not recommended, but the decision is ultimately up to you and your requirements for accuracy as an analyst.

The spherical model approach tries to better approximate reality by avoiding the problems resulting from smashing the Earth onto a flat surface. As the name suggests, this model uses a perfect sphere for representing the Earth (similar to a physical globe), which allows us to work with degrees directly. This model ignores the fact that the Earth is really more of an egg-shaped ellipsoid with varying degrees of thickness in its crust. But by working with distance on the surface of a sphere, we can begin to measure longer distances with more accuracy. The following figure illustrates this concept:

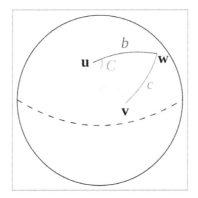

Using the ellipsoid model of the Earth, analysts strive for the best model of the Earth's surface. There are several ellipsoid models which are called datums. A **datum** is a set of values which define an estimated shape for the Earth, also known as a geodetic system. Like any other georeferencing system, a datum can be optimized for a localized area. The most commonly used datum is called WGS84, which is designed for global use. You should be aware that the WGS84 is occasionally updated as assessment techniques and technology improves. The most recent revision occurred in 2004. In North America, the NAD83 datum is used to optimize referencing over the continent. In the Eastern Hemisphere, the **European Terrestrial Reference System 1989 (ETRS89)** is used more frequently. ETRS89 is fixed to the stable part of the **Eurasian Plate**. Maps of Europe based on ETRS89 are immune to continental drift which changes up to 2.5 cm per year as the Earth's crust shifts.

An ellipsoid does not have a constant radius from the center. This fact means the formulas used in the spherical model of the Earth begin to have issues in the ellipsoid model. Though not a perfect approximation, it is much closer to reality than the spherical model. The following figure shows a generic ellipsoid model denoted by a black line contrasted against a representation of the Earth's uneven crust using the red line to represent the geoid. Although we will not use it for these examples, another model is the geoid model. The geoid is the most precise and accurate model of the Earth which is based on the Earth's surface with no influences except gravity and rotation. The following graphic is a representation of a geoid, ellipsoid, and spherical model to demonstrate the differences:

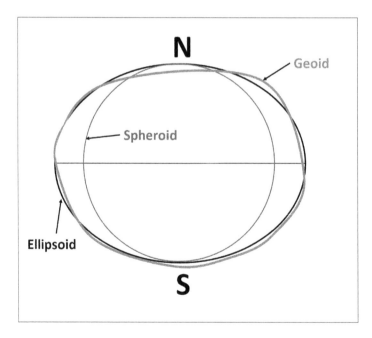

Pythagorean theorem

Now that we've discussed these different models of the Earth and the issues in measuring them, let's look at some solutions using Python. We'll start measuring with the simplest method using the Pythagorean Theorem, also known as Euclidean distance. If you remember your geometry lessons from school, the Pythagorean theorem asserts the following equation:

$a^2+b^2=c^2$

In this assertion, the variables a, b, and c are all sides of a triangle. You can solve for any one side if you know the other two. In this example, we'll start with two projected points in the **Mississippi Transverse Mercator (MSTM)** projection. The units of this projection are in meters. The x axis locations are measured from the central meridian defined by the westernmost location in the state. The y axis is defined from the NAD83 horizontal datum. The first point, defined as (x1,y1), represents **Jackson**—the state capital of Mississippi. The second point, defined as (x2,y2), represents the city of **Biloxi**, which is a coastal town, as shown in the following figure:

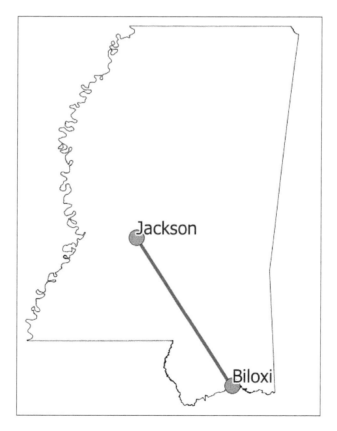

In the following example, the double asterisk (**) in Python is the syntax for exponents, which we'll use to square the distances.

We'll import the Python math module for its square root function called sqrt(). Then, we'll calculate the *x* axis and *y* axis distances. Finally, we'll use these variables to execute the Euclidean distance formula to get the distance across the bounding box in meters from an (x, y) origin used in the MSTM projection:

```
>>> import math
>>> x1 = 456456.23
>>> y1 = 1279721.064
>>> x2 = 576628.34
>>> y2 = 1071740.33
>>> x_dist = x1 - x2
>>> y_dist = y1 - y2
>>> dist_sq = x_dist**2 + y_dist**2
>>> distance = math.sqrt(dist_sq)
>>> distance
240202.66
```

So, the distance is approximately 240,202 meters, which is around 240.2 kilometers or 150 miles. This calculation is reasonably accurate because this projection is optimized for measuring distance and area in Mississippi using Cartesian coordinates.

We can also measure distance using decimal degrees, but we must perform a few additional steps. In order to measure using degrees, we must first convert the angles to radians, which accounts for the curved surface distance between the coordinates. We'll also multiply our output in radians with the radius of the Earth in meters to convert back from radians. You can read more about radians at http://en.wikipedia.org/wiki/Radian.

We'll perform this conversion using the Python math.radians() method when we calculate the x and y distances, as shown here:

```
>>> import math
>>> x1 = -90.21
>>> y1 = 32.31
>>> x2 = -88.95
>>> y2 = 30.43
>>> x_dist = math.radians(x1 - x2)
>>> y_dist = math.radians(y1 - y2)
>>> dist_sq = x_dist**2 + y_dist**2
>>> dist_rad = math.sqrt(dist_sq)
>>> dist_rad * 6371251.46
251664.46
```

OK, this time we came up with around 251 kilometers which is 11 kilometers more than our first measurement. So, as you can see, your choice of measurement algorithm and Earth model can have significant consequences. Using the same equation, we come up with radically different answers, depending on our choice of coordinate system and Earth model.

 You can read more about Euclidean distance at `http://mathworld.wolfram.com/Distance.html`.

Haversine formula

A part of the problem with just *plugging in* unprojected decimal degrees into the Pythagorean theorem is the concept of **Great Circle** distance. A Great Circle is the shortest distance between two points on a sphere. Another important feature which defines a *Great Circle* is the circle, which if followed all the way around the sphere, will bisect the sphere into two equal halves, as shown in the following Wikipedia figure (*Jhbdel, Wikipedia*):

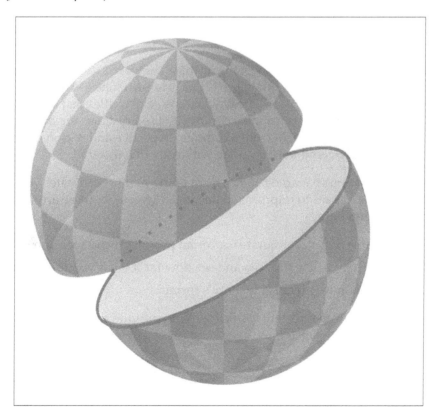

So what is the right way to measure in decimal degrees? The most popular method is the **Haversine formula** which uses trigonometry to calculate the Great Circle distance using coordinates defined in decimal degrees as input. Once again, we'll convert the axis distances from degrees to radians before we apply the formula, just like the previous example. But this time, we'll also convert the latitude (*y* axis) coordinates to radians separately, as shown here:

```
>>> import math
>>> x1 = -90.212452861859035
>>> y1 = 32.316272202663704
>>> x2 = -88.952170968942525
>>> y2 = 30.438559624660321
>>> x_dist = math.radians(x1 - x2)
>>> y_dist = math.radians(y1 - y2)
>>> y1_rad = math.radians(y1)
>>> y2_rad = math.radians(y2)
>>> a = math.sin(y_dist/2)**2 + math.sin(x_dist/2)**2 \
>>>     * math.cos(y1_rad) * math.cos(y2_rad)
>>> c = 2 * math.asin(math.sqrt(a))
>>> distance = c * 6371  # kilometers
>>> print(distance)
240.63
```

Wow! 240.6 kilometers using the Haversine formula compared to 240.2 kilometers using the optimized and more accurate projection. This difference is less than half a kilometer, which is not bad for a distance calculation of two cities that are 150 miles apart. The Haversine formula is the most commonly used distance measuring formula because it is relatively lightweight from a coding perspective and reasonably accurate in most cases. It is considered to be accurate within about a meter.

To summarize what you've learned so far, most of the point coordinates you encounter as an analyst are in unprojected decimal degrees. So, these are some of the options for measurement:

- Reproject to a distance-accurate Cartesian projection and measure
- Just use the Haversine formula and see how far it takes you for your analysis
- Use the even more precise **Vincenty's formula**

That's right! There's another formula which seeks to provide an even better measurement than Haversine.

Vincenty's formula

So we've examined distance measurement using the Pythagorean theorem (flat Earth model) and the Haversine formula (spherical Earth model). Vincenty's formula accounts for the ellipsoid model of the Earth. And if you are using a localized ellipsoid, it can be accurate within far less than a meter. In the following implementation of this formula, you can change the semi-major axis value and flattening ratio to fit the definition of any ellipsoid. Let's see what the distance is when we measure using the Vincenty's formula on the NAD83 ellipsoid:

```
import math
distance = None
x1 = -90.212452861859035
y1 = 32.316272202663704
x2 = -88.952170968942525
y2 = 30.438559624660321
# Ellipsoid Parameters
# Example is NAD83
a = 6378137  # semi-major axis
f = 1/298.257222101  # inverse flattening
b = abs((f*a)-a)  # semi-minor axis
L = math.radians(x2-x1)
U1 = math.atan((1-f) * math.tan(math.radians(y1)))
U2 = math.atan((1-f) * math.tan(math.radians(y2)))
sinU1 = math.sin(U1)
cosU1 = math.cos(U1)
sinU2 = math.sin(U2)
cosU2 = math.cos(U2)
lam = L
for i in range(100):
    sinLam = math.sin(lam)
    cosLam = math.cos(lam)
    sinSigma = math.sqrt((cosU2*sinLam)**2 +
        (cosU1*sinU2-sinU1*cosU2*cosLam)**2)
    if (sinSigma == 0):
        distance = 0  # coincident points
        break
    cosSigma = sinU1*sinU2 + cosU1*cosU2*cosLam
    sigma = math.atan2(sinSigma, cosSigma)
    sinAlpha = cosU1 * cosU2 * sinLam / sinSigma
    cosSqAlpha = 1 - sinAlpha**2
    cos2SigmaM = cosSigma - 2*sinU1*sinU2/cosSqAlpha
```

```
    if math.isnan(cos2SigmaM):
        cos2SigmaM = 0  # equatorial line
    C = f/16*cosSqAlpha*(4+f*(4-3*cosSqAlpha))
    LP = lam
    lam = L + (1-C) * f * sinAlpha * \
        (sigma + C*sinSigma*(cos2SigmaM+C*cosSigma *
        (-1+2*cos2SigmaM*cos2SigmaM)))
    if not abs(lam-LP) > 1e-12:
        break
uSq = cosSqAlpha * (a**2 - b**2) / b**2
A = 1 + uSq/16384*(4096+uSq*(-768+uSq*(320-175*uSq)))
B = uSq/1024 * (256+uSq*(-128+uSq*(74-47*uSq)))
deltaSigma = B*sinSigma*(cos2SigmaM+B/4 *
    (cosSigma*(-1+2*cos2SigmaM*cos2SigmaM) -
    B/6*cos2SigmaM*(-3+4*sinSigma*sinSigma) *
    (-3+4*cos2SigmaM*cos2SigmaM)))
s = b*A*(sigma-deltaSigma)
distance = s
print(distance)
240237.66693880095
```

Using the Vincenty's formula, our measurement came to 240.1 kilometers, which was only 100 meters off from our projected measurement using Euclidean distance. That's impressive! While it's many times more mathematically complex than the Haversine formula, you can see that it is also much more accurate.

> The pure Python `geopy` module includes an implementation of the Vincenty's formula and has the ability to geocode locations as well, by turning place names into latitude and longitude coordinates, as shown here:
>
> `http://geopy.readthedocs.org/en/latest/`

The points used in these examples are reasonably close to the equator. As you move towards the poles or work with larger or extremely small distances, the choices you make become increasingly more important. If you're just trying to make a radius around a city to select locations for a marketing campaign promoting a concert, then an error of a few kilometers is probably acceptable. However, if you're trying to estimate the volume of fuel required for an airplane to make a flight between two airports, then you want to be spot on!

If you'd like to learn more about issues with measuring distance and direction, and how to work around them with programming, visit the following site:

`http://www.movable-type.co.uk/scripts/latlong.html`

On this site, *Chris Veness* goes into great detail on this topic and provides online calculators as well as examples written in JavaScript, which are easily ported to Python. The Vincenty's formula implementation that we just saw is ported from the JavaScript on this site.

Calculating line direction

In addition to distance, you'll often want to know the bearing of a line between its end points. We can calculate this line direction from one of the points using only the Python **math** module, as shown in the following calculation:

```
>>> from math import atan2, cos, sin, degrees
>>> lon1 = -90.21
>>> lat1 = 32.31
>>> lon2 = -88.95
>>> lat2 = 30.43
>>> angle = atan2(cos(lat1)*sin(lat2)-sin(lat1) *
>>>     cos(lat2)*cos(lon2-lon1), sin(lon2-lon1)*cos(lat2))
>>> bearing = (degrees(angle) + 360) % 360
>>> print(bearing)
309.3672990606595
```

Sometimes, you end up with a negative bearing value. To avoid this issue, we add 360 to the result to avoid a negative number and use the Python module operator to keep the value from climbing over 360.

The math in the `angle` calculation is reverse engineering a right triangle and then figuring out the acute angle of the triangle. The following URL provides an explanation of the elements of this formula with an interactive example at the end:

```
https://www.mathsisfun.com/sine-cosine-tangent.html
```

Coordinate conversion

When you start working with multiple datasets you'll inevitably end up with data in different coordinate systems and projections. You can convert back and forth between **Universal Transverse Mercator (UTM)** and latitude/longitude using a pure Python module called `utm`. You can install it using `easy_install` or `pip` from PyPI available at the following link:

```
https://pypi.python.org/pypi/utm
```

The `utm` module is straightforward to use. To convert from UTM to latitude and longitude, use the following commands:

```
>>> import utm
>>> y = 479747.0453210057
>>> x = 5377685.825323031
>>> zone = 32
>>> band = 'U'
>>> print(utm.to_latlon(y, x, zone, band))
>>> (48.55199390882121, 8.725555729071763)
```

The UTM zones are numbered horizontally. However, vertically, the bands of latitude are ordered by English alphabets with a few exceptions. The letters A, B, Y, and Z cover the poles. The letters I and O are omitted because they look too much like 1 and 0. Letters N through X are in the northern hemisphere, while C through M are in the southern hemisphere. The following figure from the website, *Atlas Florae Europaeae*, illustrates the UTM zones over Europe:

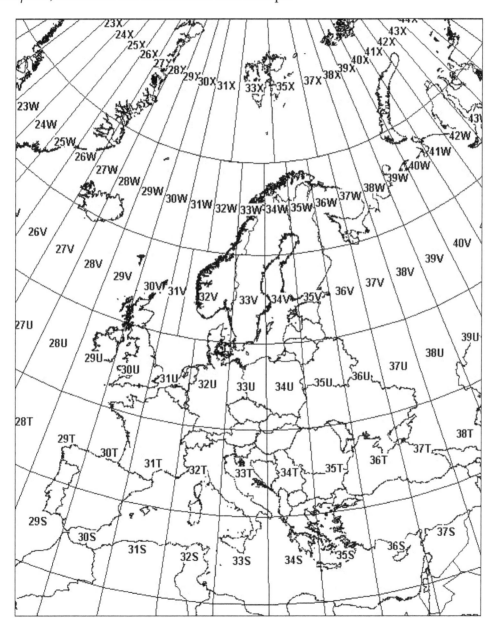

Converting from latitude and longitude is even easier. We just pass the latitude and longitude to the `from_latlon()` method, which returns a tuple with the same parameters accepted by the `to_latlon()` method:

```
>>> import utm
>>> utm.from_latlon(48.55199390882121, 8.725555729071763)
(479747.04524576373, 5377691.373080335, 32, 'U')
```

The algorithms used in this Python implementation are described in detail and are available at the following link:

```
http://www.uwgb.edu/dutchs/UsefulData/UTMFormulas.HTM
```

Reprojection

While reprojection is less common these days because of more advanced methods of data distribution, sometimes you do need to reproject a shapefile. The pure Python `utm` module works for reference system conversion, but for a full reprojection we need some help from the OGR Python API. The OGR API contained in the `osgeo` module has the Open Spatial Reference module, also known as `osr`, which we'll use for reprojection.

As an example, we'll use a `point` shapefile containing New York City museum and gallery locations in the Lambert conformal projection. We'll reproject it to WGS84 geographic (or unproject it rather). You can download this zipped shapefile at `http://git.io/vLbT4`

The following minimalist script reprojects the shapefile. The geometry is transformed and written to the new file, but the `dbf` file is simply copied to the new name as we aren't changing it. The standard Python `shutil` module, which is the shortened form for shell utilities, is used to copy `dbf`. The source and target shapefile names are variables at the beginning of the script. The target projection is also near the top, which is set using an **European Petroleum Survey Group (EPSG)** code. The script assumes there is a `.prj` projection file, which defines the source projection. If it's not, you could manually define it using the same syntax as the target projection. Each section is marked with comments, as you can see here:

```
from osgeo import ogr
from osgeo import osr
import os
import shutil
srcName = "NYC_MUSEUMS_LAMBERT.shp"
tgtName = "NYC_MUSEUMS_GEO.shp"
```

```
tgt_spatRef = osr.SpatialReference()
tgt_spatRef.ImportFromEPSG(4326)
driver = ogr.GetDriverByName("ESRI Shapefile")
src = driver.Open(srcName, 0)
srcLyr = src.GetLayer()
src_spatRef = srcLyr.GetSpatialRef()
if os.path.exists(tgtName):
    driver.DeleteDataSource(tgtName)
tgt = driver.CreateDataSource(tgtName)
lyrName = os.path.splitext(tgtName)[0]
# Use well-known binary format (WKB) to specify geometry
tgtLyr = tgt.CreateLayer(lyrName, geom_type=ogr.wkbPoint)
featDef = srcLyr.GetLayerDefn()
trans = osr.CoordinateTransformation(src_spatRef, tgt_spatRef)
srcFeat = srcLyr.GetNextFeature()
while srcFeat:
    geom = srcFeat.GetGeometryRef()
    geom.Transform(trans)
    feature = ogr.Feature(featDef)
    feature.SetGeometry(geom)
    tgtLyr.CreateFeature(feature)
    feature.Destroy()
    srcFeat.Destroy()
    srcFeat = srcLyr.GetNextFeature()
src.Destroy()
tgt.Destroy()
# Convert geometry to Esri flavor of Well-Known Text (WKT) format
# for export to the projection (prj) file.
tgt_spatRef.MorphToESRI()
prj = open(lyrName + ".prj", "w")
prj.write(tgt_spatRef.ExportToWkt())
prj.close()
srcDbf = os.path.splitext(srcName)[0] + ".dbf"
tgtDbf = lyrName + ".dbf"
shutil.copyfile(srcDbf, tgtDbf)
```

The following figure shows the reprojected points in QGIS with satellite imagery in the background:

 If you are working with a set of points, you can reproject them programmatically instead of reprojecting a shapefile using **PyProj** as given in the following link:

`http://jswhit.github.io/pyproj/`

Editing shapefiles

Shapefiles are a fundamental data format in GIS used both for exchanging data as well as performing GIS analysis. In this section, we'll learn how to work with these files extensively. In *Chapter 2, Geospatial Data*, we discussed shapefiles as a format which can have many different file types associated with it. For editing shapefiles and most other operations, we are only concerned with two types: the `.shp` file and the `.dbf` file.

The `.shp` file contains the geometry while the `.dbf` file contains the attributes of the corresponding geometry. For each geometry record in a shapefile, there is one `dbf` record. The records aren't numbered or identified in any way. This means while adding and deleting information from a shapefile, you must be careful to remove or add a record to each file type to match.

As discussed in *Chapter 4, Geospatial Python Toolbox*, there are two libraries to edit shapefiles in Python. One is the Python bindings to the OGR library. The other is the `PyShp` library, which is written in pure Python. We'll use PyShp in sticking with the *pure Python when possible* theme of this book. To install `PyShp`, use `easy_install` or `pip`.

To begin editing shapefiles, we'll start with a point shapefile containing cities for the state of Mississippi, which you can download as a ZIP file. Download the following file to your working directory and unzip it:

```
http://git.io/vLbU4
```

The points we are working with can be seen in the following figure:

Accessing the shapefile

Let's use `PyShp` to open this shapefile:

```
>>> import shapefile
>>> r = shapefile.Reader("MSCities_Geo_Pts")
>>> r
<shapefile.Reader instance at 0x00BCB760>
```

We created a shapefile `Reader` object instance and set it to the variable `r`. Notice that when we passed the filename to the `Reader` class, we didn't use any file extensions. Remember that we are dealing with at least two different files ending in `.shp` and `.dbf`. So, the base filename without the extension that is common to these two files is all we really need.

You can, however, use a file extension. PyShp will just ignore it and use the base filename. So why would you add an extension? Most operating systems allow an arbitrary number of periods in a filename. For example, you might have a shapefile with the following base name as `myShapefile.version.1.2`.

In this case, PyShp will try to interpret the characters after the last period as a file extension which would be `.2`. This issue will prevent you from opening the shapefile. So, if your shapefile has periods in the base name, you would need to add a file extension such as `.shp` or `.dbf` to the filename.

Once you have opened a shapefile and created a `Reader` object, you can get some information about the geographic data. In the following sample, we'll get the bounding box, shape type, and the number of records in the shapefile from our `Reader` object:

```
>>> r.bbox
[-91.38804855553174, 30.29314882296931, -88.18631833931401,
  34.96091138678437]
>>> r.shapeType
1
>>> r.numRecords
298
```

The bounding box, stored in the `bbox` property, is returned as a list containing the minimum x value, minimum y value, maximum x value, and maximum y value. The shape type, available as the `shapeType` property, is a numeric code defined by the official shapefile specification. In this case, 1 represents a point shapefile, 3 represents lines, and 5 represents polygons. And finally, the property `numRecords` tells us that there are 298 records in this shapefile. Because it is a simple `point` shapefile, we then know that there are 298 points, each with its own `dbf` record.

Reading shapefile attributes

The dbf file is a simple database format which is structured in a way similar to a spreadsheet with rows and columns, each column as a label defining what information it contains. We can view that information by checking the `fields` property of the `Reader` object:

```
>>> r.fields
[('DeletionFlag', 'C', 1, 0), ['STATEFP10', 'C', 2, 0],
  ['PLACEFP10', 'C', 5, 0], ['PLACENS10', 'C', 8, 0], ['GEOID10',
  'C', 7, 0], ['NAME10', 'C', 100, 0], ['NAMELSAD10', 'C', 100,
  0], ['LSAD10', 'C', 2, 0], ['CLASSFP10', 'C', 2, 0],
  ['PCICBSA10', 'C', 1, 0], ['PCINECTA10', 'C', 1, 0], ['MTFCC10',
  'C', 5, 0], ['FUNCSTAT10', 'C', 1, 0], ['ALAND10', 'N', 14, 0],
  ['AWATER10', 'N', 14,0], ['INTPTLAT10', 'C', 11, 0],
  ['INTPTLON10', 'C', 12, 0]]
```

The `fields` property returns quite a bit of information. The fields are a list with information about each field called **field descriptors**. For each field, the following information is presented:

- **Field name**: This is the name of the field as text which can be no longer than 10 characters for shapefiles.

- **Field type**: This is the type of the field which can be text, number, date, floating point number, or Boolean represented as C, N, D, F, and L respectively. The shapefile specification says it uses the dbf format specified as dBASE III, but most GIS software seems to support dBASE IV. In version IV (4), the number and floating point types are equivalent.

- **Field length**: This is the length of the data in characters or digits.

- **Decimal length**: This is the number of decimal places in a number or floating point field.

The first field descriptor outlines a hidden field, which is a part of the dbf file format specification. `DeletionFlag` allows software to mark records for deletion without actually deleting them. That way the information is still in the file but can be removed from the displayed record list or search queries.

If we just want the field name and not the other metadata, we can use Python list comprehensions to return just the first item in the descriptor and also ignore the DeletionFlag field. This example creates a list comprehension that returns the first item in each descriptor (field name) starting with the second descriptor to ignore the deletion flag:

```
>>> [item[0] for item in r.fields[1:]]
['STATEFP10', 'PLACEFP10', 'PLACENS10', 'GEOID10', 'NAME10',
    'NAMELSAD10', 'LSAD10', 'CLASSFP10', 'PCICBSA10', 'PCINECTA10',
    'MTFCC10', 'FUNCSTAT10', 'ALAND10',
'AWATER10', 'INTPTLAT10', 'INTPTLON10']
```

Now, we have just the field names which are much easier to read. For clarity, the field names all contain the number 10 because this is the version 2010 of this shapefile, which is created as a part of each census. These kinds of abbreviations are common in shapefile dbf files due to the 10 character limit on the field names.

Next, let's examine some of the records, which these fields describe. We can view an individual record using the r.record() method. We know from the first example that there are 298 records. So let's examine the third record as an example. The records are accessed using list indexes. In Python, indexes start at 0, so we have to subtract one from the desired record number to get the index. For record 3, the index would be 2. You just pass the index to the record() method:

```
>>> r.record(2)
['28', '16620', '02406337', '2816620', 'Crosby', 'Crosby town',
    '43', 'C1', 'N','N', 'G4110', 'A', 5489412, 21336,
    '+31.2742552', '-091.0614840']
```

As you can see, the field names are stored separately from the actual records. If you want to select a record value, you need its index. The index of the city name in each record is 4:

```
>>> r.record(2)[4]
'Crosby'
```

But counting indexes is tedious. It's much easier to reference a value by the field name. There are several ways we can associate a field name with the value of a particular record. The first is to use the index() method in Python lists to programmatically get the index using the field name:

```
>>> fieldNames = [item[0] for item in r.fields[1:]]
>>> name10 = fieldNames.index("NAME10")
>>> name10
4
>>> r.record(2)[name10]
'Crosby'
```

Another way we can associate field names to values is by using the Python's built-in
`zip()` method, which matches corresponding items in two or more lists and merges
them into a list of tuples. Then we can loop through that list and check the name and
then grab the associated value, as shown in the following lines of code:

```
>>> fieldNames = [item[0] for item in r.fields[1:]]
>>> fieldNames
['STATEFP10', 'PLACEFP10', 'PLACENS10', 'GEOID10', 'NAME10',
  'NAMELSAD10', 'LSAD10', 'CLASSFP10', 'PCICBSA10', 'PCINECTA10',
  'MTFCC10','FUNCSTAT10', 'ALAND10','AWATER10', 'INTPTLAT10',
  'INTPTLON10']
>>> rec = r.record(2)
>>> rec
['28', '16620', '02406337', '2816620', 'Crosby', 'Crosby town',
  '43', 'C1', 'N','N', 'G4110', 'A', 5489412, 21336,
  '+31.2742552', '-091.0614840']
>>> zipRec = zip(fieldNames, rec)
>>> list(zipRec)
[('STATEFP10', '28'), ('PLACEFP10', '16620'), ('PLACENS10',
  '02406337'), ('GEOID10', '2816620'), ('NAME10', 'Crosby'),
  ('NAMELSAD10', 'Crosby town'), ('LSAD10', '43'), ('CLASSFP10',
  'C1'), ('PCICBSA10','N'),('PCINECTA10','N'), ('MTFCC10',
  'G4110'), ('FUNCSTAT10', 'A'), ('ALAND10', 5489412),('AWATER10',
  21336), ('INTPTLAT10', '+31.2742552'), ('INTPTLON10',
  '-091.0614840')]
>>> for z in zipRec:
...     if z[0] == "NAME10": print(z[1])
...
Crosby
```

We can also loop through `dbf` records using the `r.records()` method. In this
example, we'll loop through the list returned by the `records()` method but limit the
results using Python array slicing to the first three records. As mentioned previously,
shapefiles don't contain record numbers, so we'll also enumerate the records list
and create a record number on the fly so that the output is a little easier to read.
In the next example, we'll use the `enumerate()` method, which will return tuples
containing the index and the record:

```
>>> for rec in enumerate(r.records()[:3]):
...     print(rec[0]+1, ": ", rec[1])
...
1 :  ['28', '59560', '02404554', '2859560', 'Port Gibson', 'Port
  Gibson city', '
25', 'C1', 'N', 'N', 'G4110', 'A', 4550230, 0, '+31.9558031',
  '-090.9834329']
```

```
 2 :  ['28', '50440', '02404351', '2850440', 'Natchez',
    city', '25', 'C1', 'Y', 'N', 'G4110', 'A', 34175943, 1691489,
    '+31.5495016', '-091.3887298']
 3 :  ['28', '16620', '02406337', '2816620', 'Crosby', 'Crosby
    town', '43', 'C1','N', 'N', 'G4110', 'A', 5489412, 21336,
    '+31.2742552', '-091.0614840']
```

This kind of enumeration trick is what most GIS software packages use while displaying records in a table. Many GIS analysts assume that shapefiles store the record number because every GIS program displays one. But, if you delete a record (for example, record number 5 in ArcGIS or QGIS) and save the file, on opening it again, you'll find what was formerly record number 6 is now record 5. Some spatial databases may assign a unique identifier to records. Many times a unique identifier is helpful. You can always create another field and column in dbf and assign your own number, which remains constant even when records are deleted.

If you are working with very large shapefiles, PyShp has iterator methods which access data more efficiently. The default records() method reads all records into the RAM at once, which is fine for small dbf files but becomes difficult to manage with a few thousand records. Any time you use the records() method, you can also use the r.iterRecords() method the same way. This method holds the minimum amount of information needed to provide the record at hand rather than the whole dataset. In this quick example, we use the iterRecords() method to count the number of records in order to verify the count in the file header:

```
>>> counter = 0
>>> for rec in r.iterRecords():
...     counter += 1
>>> counter
298
```

Reading shapefile geometry

Now let's take a look at the geometry. Earlier we looked at the header information and determined that this shapefile was a point shapefile. So, we know that each record contains a single point. Let's examine the first geometry record:

```
>>> geom = r.shape(0)
>>> geom.points
[[-90.98343326763826, 31.9558035947602]]
```

In each geometry record, also known as shape, the points are stored in a list called points, even if there is only one point, as in this case. Points are stored as x, y pairs, so longitude comes before latitude if that coordinate system is used.

The shapefile specification also allows for three-dimensional shapes. Elevation values are along the z axis and are often called z values. So, a three-dimensional point is typically described as x, y, z. In the shapefile format, z values are stored in a separate z attribute if allowed by the shape type. If the shape type doesn't allow for z values, then that attribute is never set when the records are read by PyShp. Shapefiles with z values also contain measure values or m values. A measure is a user assigned value which may be associated with a shape. An example would be a temperature recorded at a given location. There is another class of shape types which allow for adding m values to each shape but not z values. This class of shape types is called an **M shape type**. Just like the z values, if the data is present, the m attribute is created, otherwise it's not. You don't typically run into shapefiles with z values, and you almost never come across shapefiles with m values set. But sometimes you do, so it's good to be aware of them. And just like our fields and records in the dbf example, if you don't like having the z and m values stored in separate lists, you can use the zip() method from the points list to combine them. The zip method can take multiple lists as parameters separated by commas.

Changing a shapefile

When you create a Reader object with PyShp, it is read-only. You can change any values in the Reader object, but they are not written to the original shapefile. To create a shapefile, you need to also create a Writer object. You can change values in either a Reader or Writer object; they are just dynamic Python data types, but at some point you must copy the values from Reader to Writer. PyShp automatically handles all of the header information such as the bounding box and record count. You only need to worry about the geometry and attributes. You'll find that this method is much simpler than the OGR example we used earlier. However, it is also limited to UTM projections.

To demonstrate this concept, we'll read in a point shapefile with units in degrees and convert it to the UTM reference system in a Writer object before saving it. We'll use PyShp and the UTM module discussed earlier in this chapter. The shapefile we'll use is the New York City museums shapefile, which we reprojected earlier to the WGS84 geographic. You can also just download it as a ZIP file which is available at the following link:

```
http://git.io/vLd8Y
```

Look at the following code:

```
>>> import shapefile
>>> import utm
>>> r = shapefile.Reader("NYC_MUSEUMS_GEO")
>>> w = shapefile.Writer(r.shapeType)
>>> w.fields = list(r.fields)
```

```
>>> w.records.extend(r.records())
>>> for s in r.iterShapes():
...      lon,lat = s.points[0]
...      y,x,zone,band = utm.from_latlon(lat,lon)
...      w.point(x,y)
>>> w.save("NYC_MUSEUMS_UTM")
```

If you were to print out the first point of the first shape, you would see the following lines of code:

```
>>> print(w.shapes()[0].points[0])
[4506346.393408813, 583315.4566450359, 0, 0]
```

The point is returned as a list containing four numbers. The first two are the x and y values, while the last two are placeholders (in this case for elevation and measure values respectively, which are used when you write those types of shapefiles). Also, we did not write a PRJ projection file as we did in the reprojection example above. Here's a simple way to create a PRJ file using the EPSG code from `SpatialReference.org`. The zone variable in the previous example tells us that we are working in UTM Zone 18, which is EPSG code 26918, as shown here:

```
>>> import urllib.request
>>> prj = urllib.request.urlopen("http://spatialreference.org/ref/
epsg/
  26918/esriwkt/")
>>> with open("NYC_MUSEUMS_UTM", "w") as f:
>>>      f.write(str(prj.read()))
```

As another example, we can add a new feature to a shapefile. In this example, we'll add a second polygon to a shapefile representing a tropical storm. You can download the zipped shapefile for this example here:

```
http://git.io/vLdlA
```

We'll read the shapefile in and copy it to a `Writer` object; then we'll add the new polygon and write it back out with the same filename:

```
>>> import shapefile
>>> file_name = "ep202009.026_5day_pgn.shp"
>>> r = shapefile.Reader(file_name)
>>> w = shapefile.Writer(r.shapeType)
>>> w.fields = list(r.fields)
>>> w.records.extend(r.records())
>>> w._shapes.extend(r.shapes())
>>> w.poly(parts=[[[-104,24],[-104,25],[-103,25],[-103,24],
  [-104,24]]])
>>> w.record("STANLEY","TD","091022/1500","27","21","48","ep")
>>> w.save(file_name)
```

Adding fields

A very common operation on shapefiles is to add additional fields. This operation is easy, but there's one important element to remember. When you add a field, you must also loop through the records and either create an empty cell or add a value for that column. As an example, let's add a reference latitude and longitude column to the UTM version of the New York City museums shapefile. First, we'll open the shapefile and create a new `Writer` object. Next, we'll add the fields as float types with a length of 8 for the entire field and a maximum precision of 5 decimal places. Next, we'll open the geographic version of the shapefile and grab the coordinates from each record and add it to the corresponding attribute record in the UTM version's `dbf`. Finally, we'll save it over the original file, as you can see here:

```
>>> import shapefile
>>> r = shapefile.Reader("NYC_MUSEUMS_UTM")
>>> w = shapefile.Writer(r.shapeType)
>>> w.fields = list(r.fields)
>>> w.records.extend(r.records())
>>> w.field("LAT","F",8,5)
>>> w.field("LON","F",8,5)
>>> geo = shapefile.Reader("NYC_MUSEUMS_GEO")
>>> for i in range(geo.numRecords):
>>>    lon, lat = geo.shape(i).points[0]
>>>    w.records[i].extend([lat,lon])
>>> w._shapes.extend(r.shapes())
>>> w.save("NYC_MUSEUMS_UTM")
```

Merging shapefiles

Aggregating multiple related shapefiles of the same type into one larger shapefile is another very useful technique. You might be working as part of a team that divides up an area of interest and then assembles the data at the end of the day. You might aggregate data from a series of sensors out in the field such as weather stations. For this example, we'll use a set of building footprints for a county which are maintained separately in four different quadrants (northwest, northeast, southwest, and southeast). You can download these shapefiles as a single ZIP file available at the following link:

```
http://git.io/vLbUE
```

When you unzip these files, you'll see they are named by quadrant. The following script uses PyShp to merge them into a single shapefile:

```
>>> import glob
>>> import shapefile
>>> files = glob.glob("footprints_*shp")
>>> w = shapefile.Writer()
>>> r = None
>>> for f in files:
>>>     r = shapefile.Reader(f)
>>>     w._shapes.extend(r.shapes())
>>>     w.records.extend(r.records())
>>> w.fields = list(r.fields)
>>> w.save("Merged")
```

As you can see, merging a set of shapefiles is very straightforward. However, we didn't do any sanity checks to make sure the shapefiles were all of the same type, which you might want to do if this script was used for a repeated automated process instead of just a quick one-off process. Another note about this example is how we invoked the `Writer` object. In other examples, we used a numeric code to define a shape type. You can define that number directly (for example, 1 for `point` shapefiles) or call one of the PyShp constants. The constants are the type of shapefile in all caps. For example, a polygon is depicted as shown here:

```
shapefile.POLYGON
```

In this case, the value of that constant is 5. When copying data from a `Reader` to a `Writer` object, you'll notice that the shape type definition is simply referenced. Take a look at the following example:

```
r = shapefile.Reader("myShape")
w = shapefile.Writer(r.shapeType)
```

This last method makes your script more robust as the script has one less variable to be changed if you later change the script or the dataset. In the merging example, we don't have the benefit of having a `Reader` object available when we invoke `Writer`. We could open the first shapefile in the list and check its type, but that would add several more lines of code. An easier way is to just omit the shape type. If the `Writer` shape type isn't saved, PyShp will ignore it until you save the shapefile. At that time, it will check the individual header of a geometry record and determine it from that. While you can use this method in special cases, it's better to define the shape type explicitly, when you can, for clarity and just to be safe to prevent any outlier case errors. The following figure is a sample of this dataset to get a better idea of what the data looks like, as we'll be using it further:

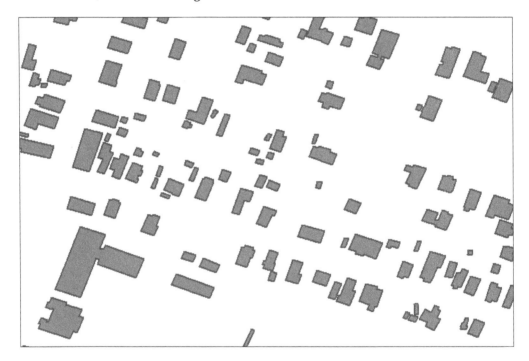

Merging shapefiles with dbfpy

The dbf portion of PyShp can occasionally run into issues with dbf files produced by certain software. Fortunately, PyShp allows you to manipulate the different shapefile types separately. A more robust dbf library named `dbfpy3` exists, which we discussed in *Chapter 4, Geospatial Python Toolbox*. You can use PyShp to handle the `shp` and `shx` files, while `dbfpy` handles more complex `dbf` files. You can download the module from this link:

```
https://github.com/GeospatialPython/dbfpy3/archive/master.zip
```

This approach takes more code, but it will often succeed where PyShp alone fails with dbf issues. The following example uses the same shapefiles used in the previous example:

```
import glob
import shapefile
from dbfpy3 import dbf
shp_files = glob.glob("footprints_*.shp")
w = shapefile.Writer()
# Loop through ONLY the shp files and copy their shapes
# to a writer object. We avoid opening the dbf files
# to prevent any field-parsing errors.
for f in shp_files:
    print("Shp: {}".format(f))
    shpf = open(f, "rb")
    r = shapefile.Reader(shp=shpf)
    w._shapes.extend(r.shapes())
    print("Num. shapes: {}".format(len(w._shapes)))
    shpf.close()
# Save only the shp and shx index file to the new
# merged shapefile.
w.saveShp("merged.shp")
w.saveShx("merged.shx")
# Now we come back with dbfpy and merge the dbf files
dbf_files = glob.glob("*.dbf")
# Use the first dbf file as a template
template = dbf_files.pop(0)
merged_dbf_name = "merged.dbf"
# Copy the entire template dbf file to the merged file
merged_dbf = open(merged_dbf_name, "wb")
temp = open(template, "rb")
merged_dbf.write(temp.read())
merged_dbf.close()
temp.close()
# Now read each record from the remaining dbf files
# and use the contents to create a new record in
# the merged dbf file.
db = dbf.Dbf(merged_dbf_name)
for f in dbf_files:
    print("Dbf: {}".format(f))
    dba = dbf.Dbf(f)
    for rec in dba:
        db_rec = db.newRecord()
        for k, v in list(rec.asDict().items()):
```

```
            db_rec[k] = v
        db_rec.store()
    db.close()
```

Splitting shapefiles

Sometimes, you may also need to split larger shapefiles to make it easier to focus on a subset of interest. This splitting or **subsetting** can be done spatially or by attributes depending on which aspect of the data is of interest.

Subsetting spatially

In the following example, we'll subset one of the quadrant files we merged. We'll filter the building footprint polygons by area and export any buildings with a 100 square meters or less (about 1,000 square feet) profile to a new shapefile. We'll use the footpints_se shapefile. PyShp has a signed area method, which accepts a list of coordinates and returns either a positive or negative area. We'll use the utm module to convert the coordinates to meters. The positive or negative area denotes whether the point order of the polygon is clockwise or counterclockwise respectively. But point order doesn't matter here, so we'll use the absolute value:

```
>>> import shapefile
>>> import utm
>>> r = shapefile.Reader("footprints_se")
>>> w = shapefile.Writer(r.shapeType)
>>> w.fields = list(r.fields)
>>> for sr in r.shapeRecords():
>>>...    utmPoints = []
>>>...    for p in sr.shape.points:
>>>...        x,y,band,zone = utm.from_latlon(p[1],p[0])
>>>...        utmPoints.append([x,y])
>>>...    area = abs(shapefile.signed_area(utmPoints))
>>>...    if  area <= 100:
>>>...        w._shapes.append(sr.shape)
>>>...        w.records.append(sr.record)
>>> w.save("footprints_185")
```

Let's see the difference in the number of records between the original and the subset:

```
>>> r = shapefile.Reader("footprints_se")
>>> subset = shapefile.Reader("footprints_185")
>>> r.numRecords
26447
>>> subset.numRecords
13331
```

We now have some substantial building blocks for geospatial analysis with vector data.

Performing selections

The previous subsetting example is one way to select data. There are many other ways to subset data for further analysis. In this section, we'll examine some of them.

Point in polygon formula

We briefly discussed the point in polygon formula in *Chapter 1*, *Learning Geospatial Analysis with Python*, as a common type of geospatial operation. You'll find that it is one of the most useful formulas out there. The formula is relatively straightforward. The following function performs this check using the **Ray Casting** method. This method draws a line from the test point all the way through the polygon and counts the number of times it crosses the polygon boundary. If the count is even, the point is outside the polygon. If it is odd, then it's inside. This particular implementation also checks to see if the point is on the edge of the polygon, as shown here:

```
def point_in_poly(x,y,poly):
    # check if point is a vertex
    if (x,y) in poly: return True
    # check if point is on a boundary
    for i in range(len(poly)):
        p1 = None
        p2 = None
        if i==0:
            p1 = poly[0]
            p2 = poly[1]
        else:
            p1 = poly[i-1]
            p2 = poly[i]
        if p1[1] == p2[1] and p1[1] == y and x > min(p1[0], \
          p2[0]) and x < max(p1[0], p2[0]):
            return True

    n = len(poly)
    inside = False

    p1x,p1y = poly[0]
    for i in range(n+1):
        p2x,p2y = poly[i % n]
        if y > min(p1y,p2y):
```

```
            if y <= max(p1y,p2y):
                if x <= max(p1x,p2x):
                    if p1y != p2y:
                        xints = (y-p1y)*(p2x-p1x)/(p2y-p1y)+p1x
                    if p1x == p2x or x <= xints:
                        inside = not inside
        p1x,p1y = p2x,p2y

    if inside: return True
    return False
```

Now, let's use the `point_in_poly()` function to test a point in Chile:

```
>>> # Test a point for inclusion
>>> myPolygon = [(-70.593016,-33.416032), (-70.589604,-33.415370),
(-70.589046,-33.417340), (-70.592351,-33.417949),
(-70.593016,-33.416032)]
>>> # Point to test
>>> lon = -70.592000
>>> lat = -33.416000
>>> print(point_in_poly(lon, lat, myPolygon))
True
```

The point is inside. Let's also verify that edge points will be detected:

```
>>> # test an edge point
>>> lon = -70.593016
>>> lat = -33.416032
>>> print(point_in_poly(lon, lat, myPolygon))
True
```

You'll find new uses for this function all the time. It's definitely one to keep in your toolbox.

Bounding Box Selections

We'll go through one more example using a simple bounding box to isolate a complex set of features and save it in a new shapefile. In this example, we'll subset the roads on the island of Puerto Rico from the mainland U.S. Major Roads shapefile. You can download the shapefile from the following link:

```
https://github.com/GeospatialPython/Learn/raw/master/roads.zip
```

Floating-point coordinate comparisons can be expensive, but we are using a box and not an irregular polygon, and so this code is efficient enough for most operations:

```
>>> import shapefile
>>> r = shapefile.Reader("roadtrl020")
>>> w = shapefile.Writer(r.shapeType)
>>> w.fields = list(r.fields)
>>> xmin = -67.5
>>> xmax = -65.0
>>> ymin = 17.8
>>> ymax = 18.6
>>> for road in r.iterShapeRecords():
>>>     geom = road.shape
>>>     rec = road.record
>>>     sxmin, symin, sxmax, symax = geom.bbox
>>>     if sxmin <  xmin: continue
>>>     elif sxmax > xmax: continue
>>>     elif symin < ymin: continue
>>>     elif symax > ymax: continue
>>>     w._shapes.append(geom)
>>>     w.records.append(rec)
>>> w.save("Puerto_Rico_Roads")
```

Attribute selections

We've now seen two different ways of subsetting a larger dataset resulting in a smaller one based on spatial relationships. Now, let's examine a quick way to subset vector data using the attribute table. In this example, we'll use a `polygon` shapefile that has densely populated urban areas within Mississippi. You can download this zipped shapefile, which is available at the following link:

```
http://git.io/vLbU9
```

This script is really quite simple. It creates the `Reader` and `Writer` objects and copies the `dbf` fields; then it loops through the records for matching attributes and adds them to `Writer`. We'll select urban areas with a population of less than 5,000, as shown here:

```
>>> import shapefile
>>> # Create a reader instance
>>> r = shapefile.Reader("MS_UrbanAnC10")
>>> # Create a writer instance
>>> w = shapefile.Writer(r.shapeType)
>>> # Copy the fields to the writer
>>> w.fields = list(r.fields)
```

```
>>> # Grab the geometry and records from all features
>>> # with the correct population
>>> selection = []
>>> for rec in enumerate(r.records()):
...     if rec[1][14] < 5000:
...         selection.append(rec)
>>> # Add the geometry and records to the writer
>>> for rec in selection:
...     w._shapes.append(r.shape(rec[0]))
...     w.records.append(rec[1])
>>> # Save the new shapefile
>>> w.save("MS_Urban_Subset")
```

Attribute selections are typically fast. Spatial selections are computationally expensive because of floating point calculations. Whenever possible, make sure you are enable to use attribute selection to subset first. The following figure shows the starting shapefile containing all urban areas on the left with a state boundary, and the urban areas with less than 5,000 people on the right after the previous attribute selection:

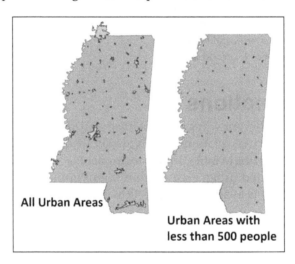

Let's see what that same example looks like using `fiona`, which takes advantage of the OGR library. We'll use nested statements to reduce the amount of code needed to properly open and close the files, as shown here:

```
>>> import fiona
>>> with fiona.open("MS_UrbanAnC10.shp") as sf:
>>>     filtered = filter(lambda f: f['properties']['POP'] < 5000,
    sf)
>>>     # Shapefile file format driver
>>>     drv = sf.driver
```

```
>>>     # Coordinate Reference System
>>>     crs = sf.crs
>>>     # Dbf schema
>>>     schm = sf.schema
>>>     subset = "MS_Urban_Fiona_Subset.shp"
>>>     with fiona.open(subset, "w",
>>>         driver=drv,
>>>         crs=crs,
>>>         schema=schm) as w:
>>>             for rec in filtered:
>>>                 w.write(rec)
```

Creating images for visualization

In *Chapter 1, Learning Geospatial Analysis with Python*, we visualized our SimpleGIS program using the Tkinter module included with Python. In *Chapter 4, Geospatial Python Toolbox*, we examined a few other methods for creating images. Now we'll examine these tools in more depth by creating two specific types of thematic maps. The first is a **dot density** map and the second is a **choropleth** map.

Dot density calculations

A dot density map shows concentrations of subjects within a given area. If an area is divided into polygons containing statistical information, you can model that information using randomly distributed dots within that area using a fixed ratio across the dataset. This type of map is commonly used for population density maps. The cat map in *Chapter 1, Learning Geospatial Analysis with Python*, is a dot density map. Let's create a dot density map from scratch using pure Python. Pure Python allows you to work with much lighter weight libraries that are generally easier to install and are more portable. For this example, we'll use a U.S. Census Bureau Tract shapefile along the U.S. Gulf Coast, which contains population data. We'll also use the point in polygon algorithm to ensure that the randomly distributed points are within the proper census tract. Finally, we'll use the PNGCanvas module to write out our image.

The PNGCanvas module is excellent and fast. However, it doesn't have the ability to fill in polygons beyond simple rectangles. You can implement a fill algorithm, but it is very slow in pure Python. However, for a quick outline and point plot, it does a great job.

You'll also see the `world2screen()` method similar to the coordinates-to-mapping algorithm we used in `SimpleGIS` in *Chapter 1, Learning Geospatial Analysis with Python,* as shown here:

```python
import shapefile
import random
import pngcanvas

def point_in_poly(x,y,poly):
    """Boolean: is a point inside a polygon?"""
    # check if point is a vertex
    if (x,y) in poly: return True
    # check if point is on a boundary
    for i in range(len(poly)):
        p1 = None
        p2 = None
        if i==0:
            p1 = poly[0]
            p2 = poly[1]
        else:
            p1 = poly[i-1]
            p2 = poly[i]
        if p1[1] == p2[1] and p1[1] == y and \
        x > min(p1[0], p2[0]) and x < max(p1[0], p2[0]):
            return True
    n = len(poly)
    inside = False
    p1x,p1y = poly[0]
    for i in range(n+1):
        p2x,p2y = poly[i % n]
        if y > min(p1y,p2y):
            if y <= max(p1y,p2y):
                if x <= max(p1x,p2x):
                    if p1y != p2y:
                        xints = (y-p1y)*(p2x-p1x)/(p2y-p1y)+p1x
                    if p1x == p2x or x <= xints:
                        inside = not inside
        p1x,p1y = p2x,p2y
    if inside: return True
    else: return False

def world2screen(bbox, w, h, x, y):
    """convert geospatial coordinates to pixels"""
    minx,miny,maxx,maxy = bbox
```

```
        xdist = maxx - minx
        ydist = maxy - miny
        xratio = w/xdist
        yratio = h/ydist
        px = int(w - ((maxx - x) * xratio))
        py = int((maxy - y) * yratio)
        return (px,py)


# Open the census shapefile
inShp = shapefile.Reader("GIS_CensusTract_poly")
# Set the output image size
iwidth = 600
iheight = 400
# Get the index of the population field
pop_index = None
dots = []
for i,f in enumerate(inShp.fields):
    if f[0] == "POPULAT11":
      # Account for deletion flag
      pop_index = i-1
# Calculate the density and plot points
for sr in inShp.shapeRecords():
    population = sr.record[pop_index]
    # Density ratio - 1 dot per 100 people
    density = population / 100
    found = 0
    # Randomly distribute points until we
    # have the correct density
    while found < density:
        minx, miny, maxx, maxy = sr.shape.bbox
        x = random.uniform(minx,maxx)
        y = random.uniform(miny,maxy)
        if point_in_poly(x,y,sr.shape.points):
            dots.append((x,y))
            found += 1
# Set up the PNG output image
c = pngcanvas.PNGCanvas(iwidth,iheight)
# Draw the red dots
c.color = (255,0,0,0xff)
for d in dots:
    # We use the *d notation to exand the (x,y) tuple
    x,y = world2screen(inShp.bbox, iwidth, iheight, *d)
    c.filled_rectangle(x-1,y-1,x+1,y+1)
# Draw the census tracts
```

```
c.color = (0,0,0,0xff)
for s in inShp.iterShapes():
    pixels = []
    for p in s.points:
        pixel = world2screen(inShp.bbox, iwidth, iheight, *p)
        pixels.append(pixel)
    c.polyline(pixels)
# Save the image
img = open("DotDensity.png","wb")
img.write(c.dump())
img.close()
```

This script outputs an outline of the census tract with the density dots to show population concentration very effectively:

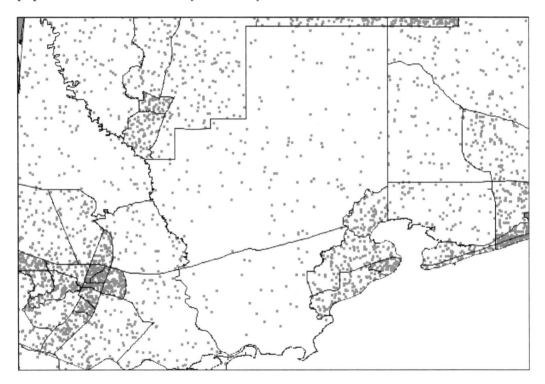

Choropleth maps

Choropleth maps also show concentration, however, they use different shades of color to show concentration. This method is useful if related data spans multiple polygons. For example, in a worldwide population density map by country, many countries have disconnected polygons (for example, Hawaii is an island state of the U.S.). In this example, we'll use the **Python Imaging Library** (**PIL**) discussed in *Chapter 3, The Geospatial Technology Landscape*. PIL is not purely Python, but it is designed specifically for Python. We'll recreate our previous dot density example as a choropleth map. We'll calculate a density ratio for each census tract based on the number of people (population) per square kilometer and use that value to adjust the color. The dark areas are more densely populated and the lighter ones are less densely populated, as shown here:

```
import math
import shapefile
try:
    import Image
    import ImageDraw
except:
    from PIL import Image, ImageDraw

def world2screen(bbox, w, h, x, y):
    """convert geospatial coordinates to pixels"""
    minx,miny,maxx,maxy = bbox
    xdist = maxx - minx
    ydist = maxy - miny
    xratio = w/xdist
    yratio = h/ydist
    px = int(w - ((maxx - x) * xratio))
    py = int((maxy - y) * yratio)
    return (px,py)

# Open our shapefile
inShp = shapefile.Reader("GIS_CensusTract_poly")
iwidth = 600
iheight = 400
# PIL Image
img = Image.new("RGB", (iwidth,iheight), (255,255,255))
# PIL Draw module for polygon fills
draw = ImageDraw.Draw(img)
# Get the population AND area index
pop_index = None
area_index = None
```

```
# Shade the census tracts
for i,f in enumerate(inShp.fields):
    if f[0] == "POPULAT11":
        # Account for deletion flag
        pop_index = i-1
    elif f[0] == "AREASQKM":
        area_index = i-1
# Draw the polygons
for sr in inShp.shapeRecords():
    density = sr.record[pop_index]/sr.record[area_index]
    # The "weight" is a scaled value to adjust the color
    # intensity based on population
    weight = min(math.sqrt(density/80.0), 1.0) * 50
    R = int(205 - weight)
    G = int(215 - weight)
    B = int(245 - weight)
    pixels = []
    for x,y in sr.shape.points:
        (px,py) = world2screen(inShp.bbox, iwidth, iheight, x, y)
        pixels.append((px,py))
    draw.polygon(pixels, outline=(255,255,255), fill=(R,G,B))
img.save("choropleth.png")
```

This script produces the following figure showing the relative density of tracks. You can adjust the color using the R, G, and B variables:

Using spreadsheets

Spreadsheets such as Microsoft Office Excel and Open Office Calc are inexpensive (even free), ubiquitous, easy to use, and great for recording structured data. For these reasons, spreadsheets are widely used to collect data for entry into a GIS format. As an analyst, you will find yourself working with spreadsheets frequently. In the previous chapters, we discussed the CSV format, which is a text file with the same basic rows and columns data structure as a spreadsheet. For CSV files, you use Python's built-in `csv` module. But most of the time, people don't bother exporting a true spreadsheet to a generic CSV file. That's where the pure Python `xlrd` module comes into play. The name `xlrd` is short for Excel Reader and is available from PyPI with its accompanying `xlwt` (Excel Writer) module. These two modules make reading and writing Excel spreadsheets a snap. Combine it with PyShp and you can move back and forth between spreadsheets and shapefiles with ease.

This example demonstrates converting a spreadsheet to a shapefile. We'll use a spreadsheet version of the New York City museums point data available at the following link:

```
https://github.com/GeospatialPython/Learn/raw/master/NYC_MUSEUMS_GEO.
xls
```

The spreadsheet contains the attribute data followed by an *x* column with the longitude and a *y* column with the latitude. To export it to a shapefile, we'll execute the following steps:

1. Open the spreadsheet.
2. Create a `shapefile Writer` object.
3. Capture the first row of the spreadsheet as the `dbf` columns.
4. Loop through each row of the spreadsheet and copy the attributes to `dbf`.
5. Create a point from the *x* and *y* spreadsheet columns.

The script for this is shown here:

```
>>> import xlrd
>>> import shapefile

>>> # Open the spreadsheet reader
>>> xls = xlrd.open_workbook("NYC_MUSEUMS_GEO.xls")
>>> sheet = xls.sheet_by_index(0)
>>>
>>> # Open the shapefile writer
>>> w = shapefile.Writer(shapefile.POINT)
>>>
```

```
>>> # Move data from spreadsheet to shapefile
>>> for i in range(sheet.ncols):
>>>    # Read the first header row
>>>    w.field(str(sheet.cell(0,i).value), "C", 40)
>>> for i in range(1, sheet.nrows):
>>>    values = []
>>>    for j in range(sheet.ncols):
>>>       values.append(sheet.cell(i,j).value)
>>>    w.record(*values)
>>>    # Pull latitude/longitude from the last two columns
>>>    w.point(float(values[-2]),float(values[-1]))
>>> w.save("NYC_MUSEUMS_XLS2SHP")
```

Converting a shapefile to a spreadsheet is a much less common operation, though not difficult. To convert a shapefile to a spreadsheet, you need to make sure you have an *x* and *y* column using the *Adding fields* example from the *Editing shapefiles* section in this chapter. You would then need to loop through the shapes and add the *x,y* values to those columns. Then, you need to read the field names and column values from dbf into an xlwt spreadsheet object or a CSV file using the csv module. The coordinate columns are labeled in the following screenshot:

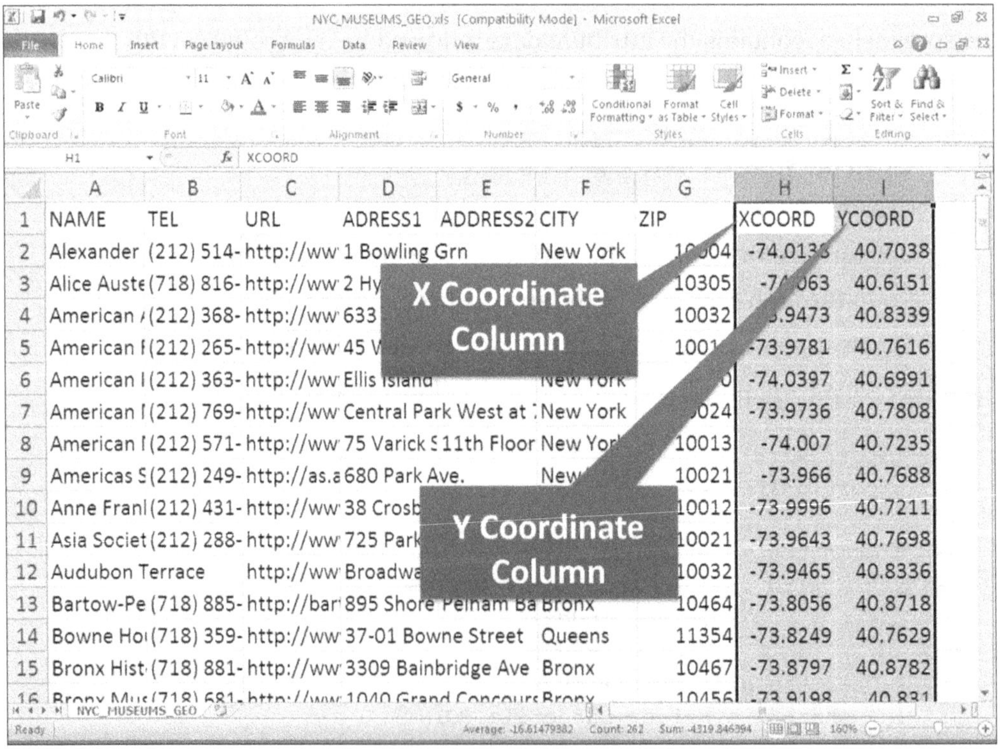

Using GPS data

The most common type of GPS data these days is the Garmin GPX format. We covered this XML format in *Chapter 4, Geospatial Python Toolbox*, which has become an unofficial industry standard. Because it is an XML format, all of the well documented rules of XML apply. However, there is another type of GPS data that pre-dates XML and GPX called **National Marine Electronics Association** (**NMEA**). These data are ASCII text sentences designed to be streamed. You occasionally bump into this format from time to time because even though it is older and esoteric, it is still very much alive and well especially to communicate ship locations via the **Automated Identification System** (**AIS**), which tracks ships globally. But as usual, you have a good option in pure Python. The `pynmea` module is available on PyPI.

Take a look at the following small sample of NMEA sentences:

```
$GPRMC,012417.859,V,1856.599,N,15145.602,W,12.0,7.27,020713,,E*4F
$GPGGA,012418.859,1856.599,N,15145.602,W,0,00,,,M,,M,,*54
$GPGLL,1856.599,N,15145.602,W,012419.859,V*35
$GPVTG,7.27,T,,M,12.0,N,22.3,K*52
$GPRMC,012421.859,V,6337.596,N,12330.817,W,66.2,23.41,020713,,E*74
```

Install the `pynmea` module from PyPI and download the complete sample file available at the following link:

```
http://git.io/vLbTv
```

Then you can run the following sample which will parse the NMEA sentences into objects. The NMEA sentences contain a wealth of information, as shown here:

```
>>> from pynmea.streamer import NMEAStream
>>> nmeaFile = open("nmea.txt")
>>> nmea_stream = NMEAStream(stream_obj=nmeaFile)
>>> next_data = nmea_stream.get_objects()
>>> nmea_objects = []
>>> while next_data:
>>>   nmea_objects += next_data
>>>   next_data = nmea_stream.get_objects()
>>> # The NMEA stream is parsed!
>>> # Let's loop through the
>>> # Python object types:
>>> for nmea_ob in nmea_objects:
>>>   if hasattr(nmea_ob, "lat"):
>>>     print("Lat/Lon: (%s, %s)" % (nmea_ob.lat, nmea_ob.lon))
```

The latitudes and longitudes are stored in a format called degrees decimal minutes. For example, 4533.35 is 45 degrees and 33.35 minutes. And .35 of a minute is exactly 21 seconds. In another example, 16708.033 is 167 degrees and 8.033 minutes. And .033 of a minute is approximately 2 seconds. You can find more information about the NMEA format at the following URL:

```
http://aprs.gids.nl/nmea/
```

Geocoding

Geocoding is the process of converting a street address to a latitude and longitude. This operation is critical to in-vehicle navigation systems and online driving direction websites. Python has two excellent geocoder libraries available named geocoder and geopy. Both take advantage of online geocoding services to allow you to geocode addresses programmatically. The geopy library even lets you reverse geocode to match a latitude and longitude to the nearest address.

First, let's go through a quick example with the geocoder library, which defaults to using Google Maps as its engine:

```
>>> import geocoder
>>> g = geocoder.google("1403 Washington Ave, New Orleans, LA
  70130")
>>> print(g.geojson)
{'type': 'Feature', 'geometry': {'type': 'Point', 'coordinates':
  [-90.08421849999999, 29.9287839]}, 'bbox': {'northeast':
  [29.9301328802915, -90.0828695197085], 'southwest':
  [29.9274349197085, -90.0855674802915]}, 'properties':
  {'quality': 'street_address', 'lat': 29.9287839, 'city': 'New
  Orleans', 'provider': 'google', 'geometry': {'type': 'Point',
  'coordinates': [-90.08421849999999, 29.9287839]}, 'lng':
  -90.08421849999999, 'method': 'geocode', 'encoding': 'utf-8',
  'confidence': 9, 'address': '1403 Washington Ave, New Orleans,
  LA 70130, USA', 'ok': True, 'neighborhood': 'Garden District',
  'county': 'Orleans Parish', 'accuracy': 'ROOFTOP', 'street':
  'Washington Ave', 'location': '1403 Washington Ave, New Orleans,
  LA 70130', 'bbox': {'northeast': [29.9301328802915,
  -90.0828695197085], 'southwest': [29.9274349197085,
  -90.0855674802915]}, 'status': 'OK', 'country': 'US', 'state':
  'LA', 'housenumber': '1403', 'postal': '70130'}}
>>> print(g.wkt)
'POINT(-90.08421849999999 29.9287839)'
```

First, we print the geojson record for that address which contains all known information in Google's database. Then we print out the returned latitude and longitude as a WKT string, which could be used as input to other operations such as checking if the address is in side of a flood plain polygon. The documentation for this library also shows you how to switch to other online geocoding services such as Bing or Yahoo! Some of these services require an API key and may have request limits.

Now let's look at the geopy library. In this example, we'll geocode against the OpenStreetMap database. Once we match the address to a location, we'll turn around and reverse geocode it, as shown here:

```
>>> from geopy.geocoders import Nominatim
>>> g = Nominatim()
>>> location = g.geocode("88360 Diamondhead Dr E, Diamondhead, MS
    39525")
>>> rev = g.reverse("{},{}".format(location.latitude,
    location.longitude))
>>> print(rev)
NVision Solutions Inc., 88360, Diamondhead Drive East,
    Diamondhead, Hancock County, Mississippi, 39520, United States
    of America
>>> print(location.raw)
{'class': 'office', 'type': 'yes', 'lat': '30.3961962', 'licence':
    'Data © OpenStreetMap contributors, ODbL 1.0.
    http://www.openstreetmap.org/copyright', 'display_name':
    'NVision Solutions Inc., 88360, Diamondhead Drive East,
    Diamondhead, Hancock County, Mississippi, 39520, United States
    of America', 'lon': '-89.3462139', 'boundingbox': ['30.3961462',
    '30.3962462', '-89.3462639', '-89.3461639'], 'osm_id':
    '2470309304', 'osm_type': 'node', 'place_id': '25470846',
    'importance': 0.421}
```

Summary

This chapter covered the critical components of GIS analysis. You examined the challenges of measuring on the curved surface of the Earth using different approaches. You looked at the basics of coordinate conversion and full reprojection using OGR, the utm module with PyShp, and Fiona which simplifies OGR. We edited shapefiles and performed spatial and attribute selections. We created thematic maps from scratch using only Python. We imported data from spreadsheets. We parsed GPS data from NMEA streams. And we used geocoding to convert street addresses to locations and back.

As a geospatial analyst, you may be familiar with both GIS and remote sensing, but most analysts specialize in one field or the other. That is why, this book approaches the fields in separate chapters in order to focus on their differences. In *Chapter 6, Python and Remote Sensing*, we'll tackle remote sensing. In GIS, we have been able to explore the field using pure Python modules. In remote sensing, we'll become more dependent on bindings to compiled modules written in C due to the sheer size and complexity of the data.

6
Python and Remote Sensing

In this chapter, we will discuss **remote sensing**. This field grows more exciting everyday as more satellites are launched and the distribution of data becomes easier. The high availability of satellite and aerial images as well as interesting new types of sensors launching each year is changing the role that remote sensing plays in understanding our world.

In this field, Python is quite capable. However, in this chapter, we will rely more on the Python bindings to the C libraries than we have in the previous chapters, where the focus was more on using pure Python. The only reason for this change is the size and complexity of remotely-sensed data. In remote sensing, we step through each pixel in an image and perform some form of query or mathematical process. An image can be thought of as a large numerical array. In remote sensing, these arrays can be quite large to the order of tens of megabytes to several gigabytes. While Python is fast, only C-based libraries can provide the speed needed to loop through arrays at a tolerable speed.

The compromise that we make in this chapter is that whenever possible, we'll use the **Python Imaging Library** (**PIL**) for image processing and NumPy, which provides multidimensional array mathematics. While written in C for speed, these libraries are designed for Python and provide a pythonic API.

In this chapter, we'll start with basic image manipulation and build on each exercise all the way to automatic change detection. The following are the topics that we'll cover:

- Swapping image bands
- Creating image histograms
- Classifying images
- Extracting features from images
- Using change detection

Swapping image bands

Our eyes can see colors only in the visible spectrum as combinations of **Red, Green, and Blue (RGB)**. Air- and space-borne sensors can collect wavelengths of the energy outside the visible spectrum. In order to view this data, we move images representing different wavelengths of light reflectance in and out of the RGB channels to make color images. These images often end up as bizarre and alien color combinations that can make visual analysis difficult. An example of a typical satellite image is seen in the following Landsat 7 satellite scene near the NASA Stennis Space Center in Mississippi along the Gulf of Mexico, which is a leading center for remote sensing and geospatial analysis in general:

Most of the vegetation appears red, and water appears almost black. This image is a type of false color image meaning that the color of the image is not based on RGB light. However, we can change the order of the bands or swap certain bands to create another type of false-color image that looks more like the world that we are used to seeing. In order to do so, you first need to download this image as a ZIP file from `http://git.io/vqs41`.

We installed the GDAL library with Python bindings in the *Installing GDAL and NumPy* sections in *Chapter 4, Geospatial Python Toolbox*. The GDAL library includes a module called `gdal_array` that loads and saves remotely-sensed images to and from the NumPy arrays for easy manipulation. GDAL itself is a data access library and does not provide much in the name of processing. So, in this chapter, we will rely heavily on NumPy to actually change images.

In this example, we'll load the image to a NumPy array using `gdal_array`, and then we'll immediately save it back to a new GeoTIFF file. However, on saving, we'll use NumPy's advanced array-slicing feature to change the order of the bands. Images in NumPy are multidimensional arrays in the order of band, height, and width. So, an image with three bands will be an array of length three, containing an array for each band, height, and width of the image. It's important to note that NumPy references the array locations as y,x (row, column) instead of the usual x, y (column, row) format that we work with in spreadsheets and other software:

```python
from osgeo import gdal_array
# name of our source image
src = "FalseColor.tif"
# load the source image into an array
arr = gdal_array.LoadFile(src)
# swap bands 1 and 2 for a natural color image.
# We will use numpy "advanced slicing" to reorder the bands.
# Using the source image
output = gdal_array.SaveArray(arr[[1, 0, 2], :], "swap.tif",
                    format="GTiff", prototype=src)
# Dereference output to avoid corrupted file on some platforms
output = None
```

In the `SaveArray` method, the last argument is called `prototype`. This argument lets you specify another image for GDAL from which to copy spatial reference information and some other image parameters. Without this argument, we'd end up with an image without georeferencing information, which could not be used in a GIS. In this case, we specified our input image file name because the images are identical except for the band order.

The result of this example produces the `swap.tif` image, which is a much more visually appealing image with green vegetation and blue water:

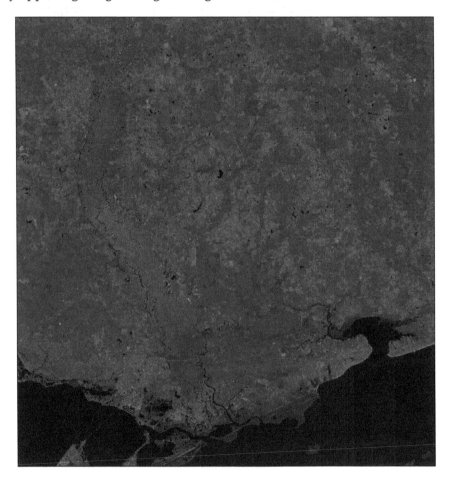

There's only one problem with this image. It's kind of dark and difficult to see. Let's see if we can figure out why.

Creating histograms

A histogram shows the statistical frequency of the data distribution in a dataset. In the case of remote sensing, the dataset is an image; the data distribution is the frequency of pixels in the range of 0 to 255, which is the range of 8-byte numbers used to store image information on computers. In an RGB image, color is represented as a three-digit tuple with (0,0,0, 0, 0) as black and (255,255,255) as white. We can graph the histogram of an image with the frequency of each value along the y axis and the range of 256 possible pixel values along the x axis.

In *Creating the simplest possible Python GIS* section of *Chapter 1*, *Learning Geospatial Analysis with Python*, we used the turtle graphics engine included with Python to create a simple GIS. Well, we can use it to easily graph histograms as well. Histograms are usually a one-off product that makes a quick script great. Additionally, histograms are typically displayed as a bar graph with the width of the bars representing the size of grouped data bins. However, in an image, each bin is only one value so we'll create a line graph. We'll use the `histogram` function in this example, and create a red, green, and blue line for each respective band. The graphing portion of this example also defaults to scaling the y axis values to the maximum RGB frequency found in the image. Technically, the y axis represents the maximum frequency that is the number of pixels in the image, which would be the case if the image was all of one color. We'll use the `turtle` module again, but this example can be easily converted into any graphical output module. Let's take a look at our `swap.tif` image:

```
from osgeo import gdal_array
import turtle as t

def histogram(a, bins=list(range(0, 256))):
    fa = a.flat
    n = gdal_array.numpy.searchsorted(gdal_array.numpy.sort(fa),
      bins)
    n = gdal_array.numpy.concatenate([n, [len(fa)]])
    hist = n[1:]-n[:-1]
    return hist

def draw_histogram(hist, scale=True):
    t.color("black")
    axes = ((-355, -200), (355, -200), (-355, -200), (-355, 250))
    t.up()
    for p in axes:
        t.goto(p)
        t.down()
    t.up()
    t.goto(0, -250)
```

```
t.write("VALUE", font=("Arial, ", 12, "bold"))
t.up()
t.goto(-400, 280)
t.write("FREQUENCY", font=("Arial, ", 12, "bold"))
x = -355
y = -200
t.up()
for i in range(1, 11):
    x = x+65
    t.goto(x, y)
    t.down()
    t.goto(x, y-10)
    t.up()
    t.goto(x, y-25)
    t.write("{}".format((i*25)), align="center")
x = -355
y = -200
t.up()
pixels = sum(hist[0])
if scale:
    max = 0
    for h in hist:
        hmax = h.max()
        if hmax > max:
            max = hmax
    pixels = max
label = int(pixels/10)
for i in range(1, 11):
    y = y+45
    t.goto(x, y)
    t.down()
    t.goto(x-10, y)
    t.up()
    t.goto(x-15, y-6)
    t.write("{}".format((i*label)), align="right")
x_ratio = 709.0 / 256
y_ratio = 450.0 / pixels
colors = ["red", "green", "blue"]
for j in range(len(hist)):
    h = hist[j]
    x = -354
    y = -199
    t.up()
    t.goto(x, y)
```

```
        t.down()
        t.color(colors[j])
    for i in range(256):
        x = i * x_ratio
        y = h[i] * y_ratio
        x = x - (709/2)
        y = y + -199
        t.goto((x, y))
im = "swap.tif"
histograms = []
arr = gdal_array.LoadFile(im)
for b in arr:
    histograms.append(histogram(b))
draw_histogram(histograms)
t.pen(shown=False)
t.done()
```

Here's what the histogram for `swap.tif` looks like after running the example:

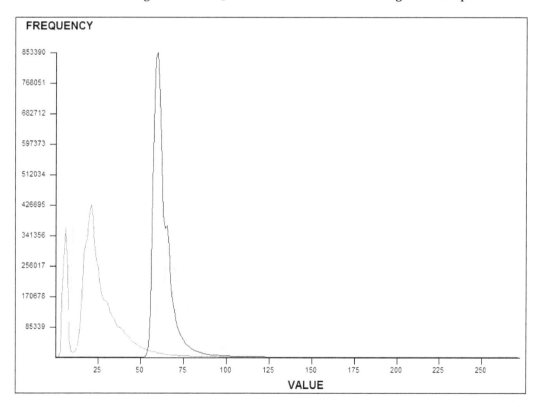

As you can see, all the three bands are grouped closely towards the left side of the graph and all have values less than 125 or so. As these values approach zero, the image becomes darker, which is not surprising. Just for fun, let's run the script again, and when we call the `draw_histogram()` function, we'll add the `scale=False` option to get a sense of the size of the image and provide an absolute scale. So, change the following line:

```
draw_histogram(histograms)
```

Replace the preceding line with the following:

```
draw_histogram(histograms, scale=False)
```

This change will produce the following histogram graph:

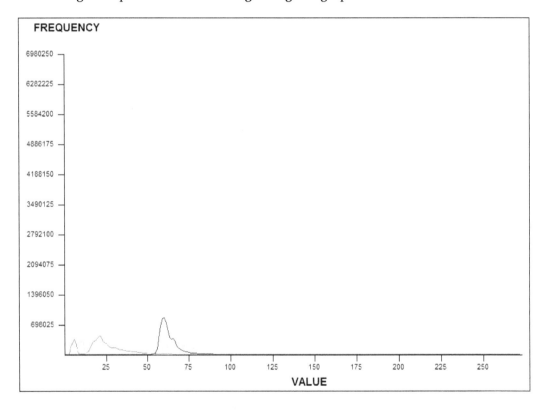

As you can see, it's difficult to see the details of the value distribution. However, this absolute-scale approach is useful if you are comparing multiple histograms from different products produced from the same source image.

So, now that we understand the basics of looking at an image statistically using histograms, how do we make our image brighter?

Performing a histogram stretch

A **histogram stretch** operation does exactly what the name says. It redistributes the pixel values across the whole scale. By doing so, we have more values at the higher-intensity level and the image becomes brighter. So, in this example, we'll reuse our histogram function, but we'll add another function called stretch() that takes an image array, creates the histogram, and then spreads out the range of values for each band. We'll run these functions on swap.tif and save the result in an image called stretched.tif:

```
import gdal_array
import operator
from functools import reduce

def histogram(a, bins=list(range(0, 256))):
    fa = a.flat
    n = gdal_array.numpy.searchsorted(gdal_array.numpy.sort(fa),
      bins)
    n = gdal_array.numpy.concatenate([n, [len(fa)]])
    hist = n[1:]-n[:-1]
    return hist

def stretch(a):
    """
    Performs a histogram stretch on a gdal_array array image.
    """
    hist = histogram(a)
    lut = []
    for b in range(0, len(hist), 256):
        # step size
        step = reduce(operator.add, hist[b:b+256]) / 255
        # create equalization lookup table
        n = 0
        for i in range(256):
            lut.append(n / step)
            n = n + hist[i+b]
    gdal_array.numpy.take(lut, a, out=a)
    return asrc = "swap.tif"
arr = gdal_array.LoadFile(src)
stretched = stretch(arr)
output = gdal_array.SaveArray(arr, "stretched.tif",
  format="GTiff", prototype=src)
output = None
```

The stretch algorithm will produce the following image. Look how much brighter and visually appealing it is!

We can run our `turtle` graphics histogram script on `stretched.tif` by changing the file name in the `im` variable to `stretched.tif`:

```
im = "stretched.tif"
```

This run will give us the following histogram:

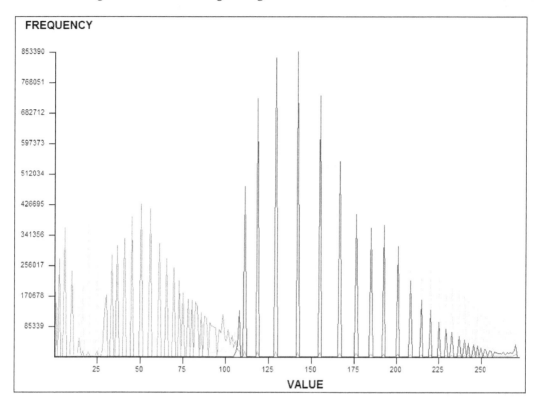

As you can see, all three bands are distributed evenly now. Their relative distribution to each other is the same, but, in the image, they are spread across the spectrum.

Clipping images

Very rarely is an analyst interested in an entire satellite scene, which can easily cover hundreds of square miles. Given the size of satellite data, we are highly motivated to reduce the size of an image to our area of interest only. The best way to accomplish this reduction is to clip an image to a boundary that defines our study area. We can use shapefiles (or other vector data) as our boundary definition and basically get rid of all the data outside this boundary. The following image contains our `stretched.tif` image with a county boundary file layered on the top, visualized in **Quantum GIS (QGIS)**:

In order to clip the image, our next example executes the following steps:

1. Load the image in an array using `gdal_array`.

2. Create a `shapefile` reader using `PyShp`.

3. Rasterize `shapefile` into a georeferenced image (convert from a vector into raster).

4. Turn the `shapefile` image into a binary mask or filter to grab only the image pixels that we want within the shapefile boundary.

5. Filter the satellite image through the mask.

6. Discard satellite image data outside the mask.

7. Save the clipped satellite image as `clip.tif`.

We installed `PyShp` in *Chapter 4, Geospatial Python Toolbox*, so you should already have it installed from PyPI. We also add a couple of useful new utility functions to this script. The first is `world2pixel()` that uses the GDAL `GeoTransform` object to do the world-coordinate to image-coordinate conversion for us. It's still the same process that we've used throughout the book, but it's better integrated with GDAL. We also add the `imageToArray()` function, which converts a PIL image to a NumPy array. The county boundary shapefile is the `hancock.shp` boundary that we've used in the previous chapters, but you can also download it here:

`http://git.io/vqsRH`

We use PIL because it is the easiest way to rasterize our shapefile as a mask image to filter out the pixels beyond the shapefile boundary:

```
import operator
from osgeo import gdal, gdal_array, osr
import shapefile
try:
    import Image
    import ImageDraw
except:
    from PIL import Image, ImageDraw

# Raster image to clip
raster = "stretched.tif"
# Polygon shapefile used to clip
shp = "hancock"
# Name of clipped raster file(s)
output = "clip"

def imageToArray(i):
    """
    Converts a Python Imaging Library array to a gdal_array image.
    """
    a = gdal_array.numpy.fromstring(i.tostring(), 'b')
    a.shape = i.im.size[1], i.im.size[0]
```

```
        return a

def world2Pixel(geoMatrix, x, y):
    """
    Uses a gdal geomatrix (gdal.GetGeoTransform()) to calculate
    the pixel location of a geospatial coordinate
    """
    ulX = geoMatrix[0]
    ulY = geoMatrix[3]
    xDist = geoMatrix[1]
    yDist = geoMatrix[5]
    rtnX = geoMatrix[2]
    rtnY = geoMatrix[4]
    pixel = int((x - ulX) / xDist)
    line = int((ulY - y) / abs(yDist))
    return (pixel, line)
# Load the source data as a gdal_array array
srcArray = gdal_array.LoadFile(raster)
# Also load as a gdal image to get geotransform (world file) info
srcImage = gdal.Open(raster)
geoTrans = srcImage.GetGeoTransform()
# Use pyshp to open the shapefile
r = shapefile.Reader("{}.shp".format(shp))
# Convert the layer extent to image pixel coordinates
minX, minY, maxX, maxY = r.bbox
ulX, ulY = world2Pixel(geoTrans, minX, maxY)
lrX, lrY = world2Pixel(geoTrans, maxX, minY)
# Calculate the pixel size of the new image
pxWidth = int(lrX - ulX)
pxHeight = int(lrY - ulY)
clip = srcArray[:, ulY:lrY, ulX:lrX]
# Create a new geomatrix for the image
# to contain georeferencing data
geoTrans = list(geoTrans)
geoTrans[0] = minX
geoTrans[3] = maxY
# Map points to pixels for drawing the county boundary
# on a blank 8-bit, black and white, mask image.
pixels = []
for p in r.shape(0).points:
    pixels.append(world2Pixel(geoTrans, p[0], p[1]))
rasterPoly = Image.new("L", (pxWidth, pxHeight), 1)
# Create a blank image in PIL to draw the polygon.
rasterize = ImageDraw.Draw(rasterPoly)
```

```
rasterize.polygon(pixels, 0)
# Convert the PIL image to a NumPy array
mask = imageToArray(rasterPoly)
# Clip the image using the mask
clip = gdal_array.numpy.choose(mask, (clip, 0)).astype(
                              gdal_array.numpy.uint8)
# Save ndvi as tiff
gdal_array.SaveArray(clip, "{}.tif".format(output),
                     format="GTiff", prototype=raster)
```

This script produces the following clipped image. The areas remaining outside the county boundary are actually called the `NoData` or `fill` values and are displayed in black, but ignored by most geospatial software. As images are rectangles, the `NoData` values are common for data that does not completely fill an image:

You have now walked through an entire workflow that is used by geospatial analysts around the world everyday to prepare multispectral satellite and aerial images for use in a GIS. Now, let's look at how we can actually analyze images as information.

Classifying images

Automated Remote Sensing (**ARS**)is rarely ever done in the visible spectrum. ARS is the processing of an image without any human input. The most commonly available wavelengths outside of the visible spectrum are infrared and near-infrared. The following scene is a thermal image (`band 10`) from a fairly recent Landsat 8 flyover of the U.S. Gulf Coast from New Orleans, Louisiana to Mobile, Alabama. The major natural features in the image are labeled so that you can orient yourself:

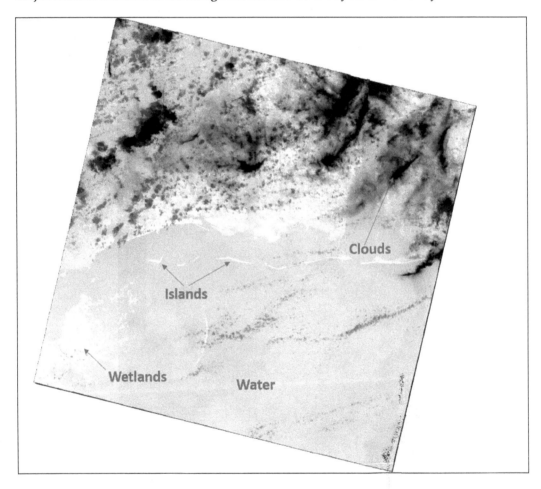

As every pixel in this image has a reflectance value, it is information. Python can *see* these values and pick out features in the same way that we intuitively do by grouping the related pixel values. We can colorize the pixels based on their relation to each other to simplify the image and view the related features. This technique is called classification.

Classification can range from fairly simple groupings based only on some value distribution algorithm derived from the histogram, to complex methods involving training datasets and even computer learning and artificial intelligence. The simplest forms are called **unsupervised classifications** in which no additional input is given other than the image itself. Methods involving some sort of training data to guide the computer are called **supervised classifications**. It should be noted that classification techniques are used across many fields, from medical doctors searching for cancerous cells in a patient's body scan to casinos using a facial recognition software on security videos to automatically spot known con artists at blackjack tables.

To introduce remote sensing classification, we'll just use the histogram to group the pixels with similar colors and intensities and see what we get. First, you'll need to download the Landsat 8 scene:

```
https://github.com/GeospatialPython/Learn/blob/master/thermal.
zip?raw=true
```

Instead of our `histogram()` function from the previous examples, we'll use the version included with NumPy that allows you to specify a number of bins easily and returns two arrays with the frequency as well as ranges of the bin values. We'll use the second array with the ranges as our class definition for the image. The `lut` or look-up table is an arbitrary color palette used to assign colors to the 20 unsupervised classes. You can use any colors that you want.

```python
"""Classify a remotely sensed image"""

# https://github.com/GeospatialPython/Learn/raw/master/thermal.zip

from osgeo import gdal_array

# Input file name (thermal image)
src = "thermal.tif"

# Output file name
tgt = "classified.jpg"

# Load the image into numpy using gdal
srcArr = gdal_array.LoadFile(src)

# Split the histogram into 20 bins as our classes
classes = gdal_array.numpy.histogram(srcArr, bins=20)[1]

# Color look-up table (LUT) - must be len(classes)+1.
# Specified as R, G, B tuples
lut = [[255, 0, 0], [191, 48, 48], [166, 0, 0], [255, 64, 64],
    [255, 115, 115],
```

```
        [255, 116, 0], [191, 113, 48], [255, 178, 115], [0, 153,
   153],
        [29, 115, 115], [0, 99, 99], [166, 75, 0], [0, 204, 0],
   [51, 204, 204],
        [255, 150, 64], [92, 204, 204], [38, 153, 38], [0, 133, 0],
        [57, 230, 57], [103, 230, 103], [184, 138, 0]]

# Starting value for classification
start = 1

# Set up the RGB color JPEG output image
rgb = gdal_array.numpy.zeros((3, srcArr.shape[0],
                             srcArr.shape[1], ),
  gdal_array.numpy.float32)

# Process all classes and assign colors
for i in range(len(classes)):
    mask = gdal_array.numpy.logical_and(start <= srcArr, srcArr <=
      classes[i])
    for j in range(len(lut[i])):
        rgb[j] = gdal_array.numpy.choose(mask, (rgb[j],
          lut[i][j]))
    start = classes[i]+1

# Save the image
output = gdal_array.SaveArray(rgb.astype(gdal_array.numpy.uint8),
  tgt, format="JPEG")
output = None
```

The following image is our classification output, which we just saved as JPEG.

We didn't specify the prototype argument when saving an image, so it has no georeferencing information, though we could easily have done otherwise to save the output as a GeoTIFF.

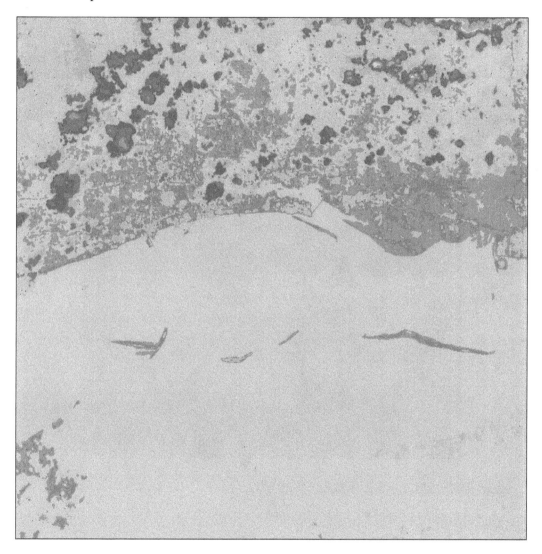

This result isn't bad for a very simple unsupervised classification. The islands and coastal flats show up as different shades of green. The clouds were isolated as shades of orange and dark blues. We did have some confusion in land where the land features were colored the same as the Gulf of Mexico. We could further refine this process by defining the class ranges manually instead of just using the histogram.

Extracting features from images

The ability to classify an image leads us to another remote sensing capability. Now that you've worked with shapefiles over the last few chapters, have you ever wondered where they come from? Vector GIS data such as shapefiles are typically extracted from remotely-sensed images like the examples that we've seen so far. Extraction normally involves an analyst clicking around each object in an image and drawing the feature to save it as data. It is also possible with good remotely-sensed data and proper preprocessing to automatically extract features from an image.

For this example, we'll take a subset of our Landsat 8 thermal image to isolate a group of barrier islands, as shown in the following screenshot:

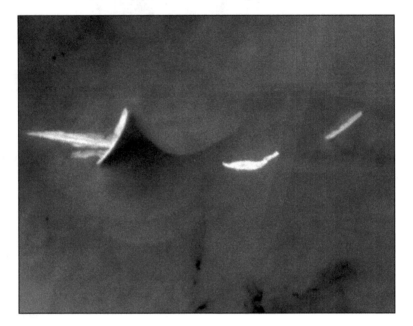

You can download this image here:

http://git.io/vqarj

Our goal with this example is to automatically extract the three islands in the image as a shapefile. Before we can do this, we need to mask out any data that we aren't interested in. For example, the water has a wide range of pixel values as do the islands themselves. If we want to extract just the islands, we need to push all the pixel values to just two bins to make the image black and white. This technique is called **thresholding**. The islands in the image have enough contrast with the water in the background such that thresholding should isolate them nicely.

In the following script, we will read the image to an array and then histogram the image using only two bins. We will then use the colors black and white to color the two bins. This script is simply a modified version of our classification script with a very limited output:

```python
from osgeo import gdal_array

# Input file name (thermal image)
src = "islands.tif"

# Output file name
tgt = "islands_classified.tiff"

# Load the image into numpy using gdal
srcArr = gdal_array.LoadFile(src)

# Split the histogram into 20 bins as our classes
classes = gdal_array.numpy.histogram(srcArr, bins=2)[1]

lut = [[255, 0, 0], [0, 0, 0], [255, 255, 255]]

# Starting value for classification
start = 1

# Set up the output image
rgb = gdal_array.numpy.zeros((3, srcArr.shape[0], srcArr.shape[1],
    ),
                              gdal_array.numpy.float32)

# Process all classes and assign colors
for i in range(len(classes)):
    mask = gdal_array.numpy.logical_and(start <= srcArr, srcArr <=
        classes[i])
    for j in range(len(lut[i])):
        rgb[j] = gdal_array.numpy.choose(mask, (rgb[j],
            lut[i][j]))
    start = classes[i]+1

# Save the image
gdal_array.SaveArray(rgb.astype(gdal_array.numpy.uint8),
                     tgt, format="GTIFF", prototype=src)
```

The output looks great, as shown in the following screenshot:

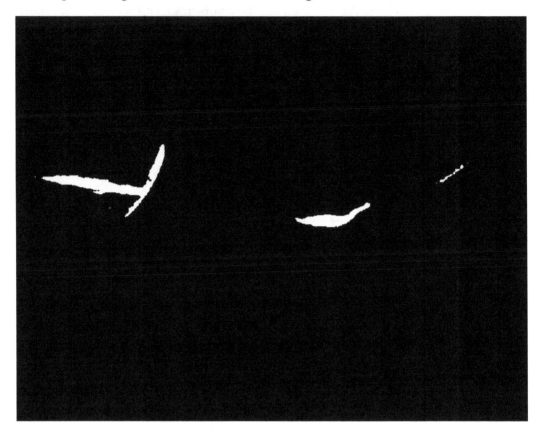

The islands are clearly isolated so that our extraction script will be able to identify them as polygons and save them to a shapefile. The GDAL library has a method called `Polygonize()` that does exactly this. It groups all the sets of isolated pixels in an image and saves them as a feature dataset. One interesting technique that we will use in this script is to use our input image as a mask. The `Polygonize()` method allows you to specify a mask that will use the color black as a filter and will prevent the water from being extracted as a polygon, so we'll end up with just the islands. Another point to note in the script is that we copy the georeferencing information from our source image to our shapefile in order to geolocate it properly:

```
from osgeo import gdal, ogr, osr

# Thresholded input raster name
src = "islands_classified.tiff"
# Output shapefile name
tgt = "extract.shp"
```

```
# OGR layer name
tgtLayer = "extract"
# Open the input raster
srcDS = gdal.Open(src)
# Grab the first band
band = srcDS.GetRasterBand(1)
# Force gdal to use the band as a mask
mask = band
# Set up the output shapefile
driver = ogr.GetDriverByName("ESRI Shapefile")
shp = driver.CreateDataSource(tgt)
# Copy the spatial reference
srs = osr.SpatialReference()
srs.ImportFromWkt(srcDS.GetProjectionRef())
layer = shp.CreateLayer(tgtLayer, srs=srs)
# Set up the dbf file
fd = ogr.FieldDefn("DN", ogr.OFTInteger)
layer.CreateField(fd)
dst_field = 0
# Automatically extract features from an image!
extract = gdal.Polygonize(band, mask, layer, dst_field, [], None)
```

The output shapefile is simply called extract.shp. In *Chapter 4*, *Geospatial Python Toolbox*, we created a quick pure Python script using PyShp and PNGCanvas to visualize shapefiles. We'll bring that script back here to look at our shapefile, but we'll add something to it. The largest island has a small lagoon that shows up as a hole in the polygon. In order to render it properly, we have to deal with the parts in a shapefile record. The previous example using this script did not do that, so we'll add that piece as we loop through the shapefile features:

```
import shapefile
import pngcanvas

r = shapefile.Reader("extract.shp")
xdist = r.bbox[2] - r.bbox[0]
ydist = r.bbox[3] - r.bbox[1]
iwidth = 800
iheight = 600
xratio = iwidth/xdist
yratio = iheight/ydist
polygons = []
for shape in r.shapes():
    for i in range(len(shape.parts)):
        pixels = []
        pt = None
```

```
        if i < len(shape.parts)-1:
            pt = shape.points[shape.parts[i]:shape.parts[i+1]]
        else:
            pt = shape.points[shape.parts[i]:]
        for x, y in pt:
            px = int(iwidth - ((r.bbox[2] - x) * xratio))
            py = int((r.bbox[3] - y) * yratio)
            pixels.append([px, py])
        polygons.append(pixels)
c = pngcanvas.PNGCanvas(iwidth, iheight)
for p in polygons:
    c.polyline(p)
f = open("extract.png", "wb")
f.write(c.dump())
f.close()
```

The following screenshot shows our automatically extracted island features. Commercial packages that do this kind of work can easily cost tens of thousands of dollars. While these packages are very robust, it is still fun and empowering to see how far you can get with simple Python scripts and a few open source packages. In many cases, you can do everything that you need to do.

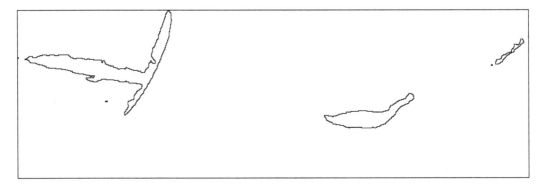

The westernmost island contains the polygon hole, as shown in the following screenshot, which is zoomed to this area:

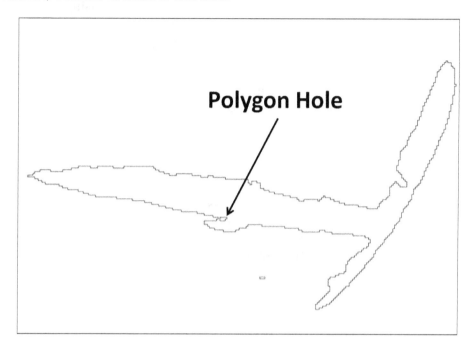

 If you want to see what would happen if we didn't deal with the polygon holes, then just run the version of the script from *Chapter 4, Geospatial Python Toolbox*, on this same shapefile to compare the difference. The lagoon is not easy to see, but you will find it if you use the other script.

Automated feature extraction is a holy grail in geospatial analysis because of the cost and tedious effort required to extract features manually. The key to feature extraction is proper image classification. Automated feature extraction works well with water bodies, islands, roads, farm fields, buildings, and other features that tend to have high-contrast pixel values in their background.

Change detection

You now have a good grasp of working with remote sensing data using GDAL, NumPy, and PIL. It's time to move on to our most complex example: **change detection**. Change detection is the process of taking two geo-registered images of the exact same area from two different dates and automatically identifying the differences. It is really just another form of image classification. It can range from trivial techniques like those used here to highly sophisticated algorithms that provide amazingly precise and accurate results.

For this example, we'll use two images from a coastal area. These images show a populated area before and after a major hurricane, so there are significant differences, many of which are easy to spot, making these samples good to learn change detection. Our technique is to simply subtract the first image from the second to get a simple image difference using NumPy. This is a valid and often used technique. The advantages are that it is comprehensive and very reliable. The disadvantages of this overly simple algorithm are that it doesn't isolate the type of change. Many changes are insignificant for analysis such as the waves in the ocean. In this example, we'll mask the water fairly effectively to avoid this distraction and only focus on the higher reflectance values toward the right side of the difference image histogram.

You can download the before image from http://git.io/vqa6h.

You can download the after image from http://git.io/vqaic.

Note that these images are quite large—24 MB and 64 MB, respectively!

These images are displayed on the following pages. The before image is panchromatic, while the after image is false color. Panchromatic images are created by sensors that capture all the visible light and are typically higher resolution than multispectral sensors that capture bands containing restricted wavelengths. Normally, you would use two identical band combinations, but these samples will work for our purpose. The visual markers that we can use to evaluate the change detection include a bridge in the southeast quadrant of the image that spans from the peninsula to the edge of the image. This bridge is clearly visible in the before image and is reduced to pilings by the hurricane. Another marker is a boat in the northwest quadrant that appears in the after image as a white trail but is not in the before image. A neutral marker is the water and the state highway, which runs through town and connects to the bridge. This feature is highly reflective concrete that is easily visible in imagery and does not change significantly between the two images. The following is a screenshot of the before image:

To view these images up close yourself, you should use QGIS or **OpenEV** (**FWTools**), described in the *Quantum GIS* and *OpenEV* sections in *Chapter 3, The Geospatial Technology Landscape*. The following image is the after image:

So, to perform a change detection, our example script will execute the following steps:

1. Read both the images to NumPy arrays with `gdal_array`.

2. Subtract the before image from the after image (*difference = after–before*).

3. Divide the image into five classes.

4. Set our color table to use black to mask the lower classes to filter water and roads because they are darker in the image.

5. Assign colors to the classes.

6. Save the image.

The script is relatively short, as follows:

```
from osgeo import gdal, gdal_array
import numpy as np

# "Before" image
im1 = "before.tif"
# "After" image
im2 = "after.tif"
# Load before and after into arrays
ar1 = gdal_array.LoadFile(im1).astype(np.int8)
ar2 = gdal_array.LoadFile(im2)[1].astype(np.int8)
# Perform a simple array difference on the images
diff = ar2 - ar1
# Set up our classification scheme to try
# and isolate significant changes
classes = np.histogram(diff, bins=5)[1]
# The color black is repeated to mask insignificant changes
lut = [[0, 0, 0], [0, 0, 0], [0, 0, 0], [0, 0, 0], [0, 255, 0],
    [255, 0, 0]]
# Starting value for classification
start = 1
# Set up the output image
rgb = np.zeros((3, diff.shape[0], diff.shape[1], ), np.int8)
# Process all classes and assign colors
for i in range(len(classes)):
    mask = np.logical_and(start <= diff, diff <= classes[i])
    for j in range(len(lut[i])):
        rgb[j] = np.choose(mask, (rgb[j], lut[i][j]))
    start = classes[i]+1
# Save the output image
output = gdal_array.SaveArray(rgb, "change.tif", format="GTiff",
    prototype=im2)
output = None
```

Here's what our initial difference image looks like:

For the most part, the green classes represent areas where something was added. The red would be a darker value where something was probably removed. We can see that the boat trail is green in the northwest quadrant. In the image, we can also see a lot of changes in vegetation as would be expected probably from seasonal differences. The bridge is an anomaly because the exposed pilings are brighter than the darker surface of the original bridge, which make them green instead of red. Concrete is a major indicator in change detection because it is very bright in sunlight and is usually a sign of new development. Conversely, if a building is torn down and the concrete removed, the difference is also easy to identify. So, our simple difference algorithm used here isn't perfect, but it could be greatly improved using thresholding, masking, better class definitions, and other techniques.

To really appreciate our change detection product, you can overlay it on the before or after image in QGIS and set the color black to be transparent, as shown in the following image:

Summary

In this chapter, we covered the foundations of remote sensing including band swapping, histograms, image classification, feature extraction, and change detection. As in the other chapters, we stayed as close to pure Python as possible, and where we compromised on this goal for the processing speed, we limited the software libraries as much as possible to keep things simple. However, if you have the tools from this chapter installed, you really have a complete remote sensing package that is limited only by your desire to learn.

The authors of GDAL have a set of Python examples, which cover some advanced topics that may be of interest:

`https://svn.osgeo.org/gdal/trunk/gdal/swig/python/samples/`

In the next chapter, we'll investigate elevation data. Elevation data doesn't fit squarely in GIS or remote sensing as it has elements of both types of processing.

7
Python and Elevation Data

Elevation data is one of the most fascinating types of geospatial data. It represents many different types of data sources and formats. It can display properties of both vector and raster data resulting in unique data products. Elevation data can be used for terrain visualization, land cover classification, hydrology modeling, transportation routing, feature extraction, and many other purposes.

You can't perform all of these options with both raster and vector data but as elevation data is three-dimensional, containing x, y, and z coordinates, you can often get more out of this data than any other type.

In this chapter, you're going to learn to read and write elevation data in both raster and vector formats. We'll also create some derivative products. The topics that we'll cover are as follows:

- Using ASCII Grid elevation data files
- Creating shaded relief images
- Creating elevation contours
- Gridding LIDAR data
- Creating a 3D mesh

ASCII Grid files

For most of this chapter, we'll use **ASCII Grid** files or **ASCIIGRID**. These files are a type of raster data usually associated with elevation data. This grid format stores data as text in equal-sized square rows and columns with a simple header. Each cell in a row/column stores a single numeric value, which can represent some feature of terrain, such as elevation, slope, or flow direction. The simplicity makes it an easy-to-use, platform-independent raster format. This format is described in the *ASCII Grids* section in *Chapter 2, Geospatial Data*.

Throughout the book, we've relied on GDAL and, to some extent, PIL to read and write geospatial raster data including the `gdal_array` module to load raster data in the NumPy arrays. ASCI Grid allows us to read and write rasters using only Python or even NumPy because it is simple, plain text.

As a reminder, some elevation datasets use image formats to store elevation data. Most image formats only support 8-bit values ranging between 0 to 255; however, some formats, including TIFF, can store larger values. Geospatial software can typically display these datasets; however, traditional image software and libraries usually do not. For simplicity in this chapter, we'll mostly stick to the ASCI Grid format for data, which is both human and machine readable as well as being widely supported.

Reading grids

NumPy has the ability to read the ASCI Grid format directly using its `loadtxt()` method designed to read arrays from text files. The first six lines consist of the header, which are not part of the array. The following lines are a sample of a grid header:

```
ncols           250
nrows           250
xllcorner       277750.0
yllcorner       6122250.0
cellsize        1.0
NODATA_value       -9999
```

Line 1 contains the number of columns in the grid, which is synonymous with the *x* axis. Line 2 represents the *y* axis described as a number of rows. Line 3 represents the *x* coordinate of the lower left corner, which is the minimum x value in meters. Line 4 is the corresponding minimum y value in the lower left corner of the grid. Line 5 is the cell size or resolution of the raster. As the cells are square, only one size value is needed, as opposed to the separate x and y resolution values in most geospatial rasters. The fifth line is the `NODATA_value`, which is a number assigned to any cell for which a value is not provided. Geospatial software ignores these cells for calculations and often allows special display settings for it, such as coloring them black or transparent. The `-9999` value is a common no data placeholder value used in the industry, which is easy to detect in software but can be arbitrarily selected. Elevation with negative values (that is, bathymetry) may have valid data at -9999 meters, for instance, and may select 9999 or other values. As long as this value is defined in the header, most software will have no problems. In some examples, we'll use the number zero; however, zero can often also be a valid data value.

The `numpy.loadtxt()` method includes an argument called `skiprows`, which allows you to specify a number of lines in the file to be skipped before reading the array values. To try this technique, you can download a sample grid file called `myGrid.asc` from `http://git.io/vYapU`.

So, for `myGrid.asc`, we would use the following code:

```
myArray  = numpy.loadtxt("myGrid.asc", skiprows=6)
```

This line results in the `myArray` variable containing a `numpy` array derived from the ASCIIGRID `myGrid.asc` file. The ASC filename extension is used by the ASCIIGRID format. This code works great but there's one problem. NumPy allows us to skip the header but not keep it. We need to keep it in order to have a spatial reference for our data to process as well as save this grid or create a new one.

To solve this problem, we'll use Python's built-in `linecache` module to grab the header. We could open the file, loop through the lines, store each one in a variable, and then close the file. However, `linecache` reduces the solution to a single line. The following line reads the first line in the file to a variable called `line1`:

```
import linecache
line1 = linecache.getline("myGrid.asc", 1)
```

In the examples in this chapter, we'll use this technique to create a simple header processor that can parse these headers to Python variables in just a few lines.

Writing grids

Writing grids in NumPy is just as easy as reading them. We use the corresponding `numpy.savetxt()` function to save a grid to a text file. The only catch is that we must build and add the six lines of header information before we dump the array to the file. This process is slightly different if you are using NumPy versions before 1.7 or after. In either case, you build the header as a string first. If you are using NumPy 1.7 or later, the `savetext()` method has an optional argument called `header`, which lets you specify a string as an argument. You can quickly check your NumPy version from the command line using the following command:

```
python -c "import numpy;print(numpy.__version__)"
1.9.2
```

The backward-compatible method is to open a file, write the header, and then dump the array. Here is a sample of the version 1.7 approach to save an array called `myArray` to an ASCII Grid file called `myGrid.asc`:

```
header = "ncols            {}\n".format(myArray.shape[1])
header += "nrows            {}\n".format(myArray.shape[0])
header += "xllcorner    277750.0\n"
header += "yllcorner    6122250.0\n"
header += "cellsize     1.0\n"
header += "NODATA_value     -9999"
numpy.savetxt("myGrid.asc", myArray, header=header, fmt="%1.2f")
```

We make use of Python format strings, which allow you to put placeholders in a string to format the Python objects to be inserted. The `{}` format variable turns whatever object you reference into a string. In this case, we are referencing the number of columns and rows in the array. In NumPy, an array has both a size and shape property. The `size` property returns an integer for the number of values in the array. The `shape` property returns a tuple with the number of rows and columns, respectively. So, in the preceding example, we use the `shape` property tuple to add the row and column counts to the header of our ASCII Grid. Notice that we also add a trailing newline character for each line (`\n`). There is no reason to change the x and y values, cell size, or no data value unless we altered them in the script. The `savetxt()` method also has a `fmt` argument, which allows you to use Python format strings to specify how the array values are written. In this case, the `%1.2f` value specifies floats with at least one number and no more than two decimal places.

The backward-compatible version for NumPy, before 1.6, builds the header string in the same way but creates the file handle first:

```
import numpy
f = open("myGrid.asc", "w")
f.write(header)
numpy.savetxt(f, myArray, fmt="%1.2f")
f.close()
```

In the examples in this chapter, we'll introduce Python's `with` statement as an approach to write files, which provides you with more graceful file management by ensuring that the files are closed properly. If any exceptions are thrown, the file is still closed cleanly:

```
with open("myGrid.asc", "w") as f:
    f.write(header)
    numpy.savetxt(f, str(myArray), fmt="%1.2f")
```

As you'll see in the upcoming examples, this ability to produce valid geospatial data files using only NumPy is quite powerful. In the next couple of examples, we'll be using an ASCIIGRID **Digital Elevation Model** (**DEM**) of a mountainous area near Vancouver, British Columbia in Canada. You can download this sample as a ZIP file at the following URL:

`http://git.io/vYwUX`

The following image is the raw DEM colorized using QGIS with a color ramp that makes the lower elevation values dark blue and higher elevation values bright red:

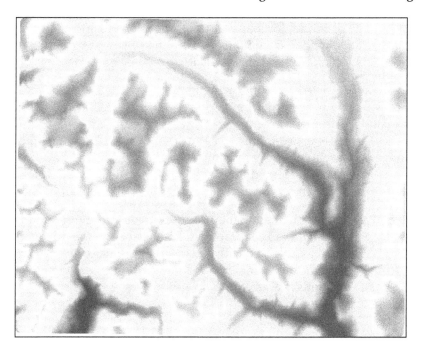

While we can conceptually understand the data in this way, it is not an intuitive way to visualize the data. Let's see if we can do better.

Creating a shaded relief

Shaded relief maps color elevation in such a way that it looks as if the terrain is cast in a low angle light, which creates bright spots and shadows. This aesthetic styling creates an almost photographic illusion, which is easy to grasp to understand the variation in the terrain. It is important to note that this style is truly an illusion as the light is often physically inaccurate in terms of the solar angle, and the elevation is usually exaggerated to increase contrast.

In this example, we'll use the ASCII DEM referenced previously to create another grid, which represents a shaded relief version of the terrain in NumPy. This terrain is quite dynamic so we won't need to exaggerate the elevation; however, the script has a variable called z, which can be increased from 1.0 to scale the elevation up.

After we define all the variables including the input and output filenames, you'll see the header parser based on the linecache module, which also uses a Python list comprehension to loop and parse the lines that are then split from a list into the six variables. We also create a y cell size called ycell, which is just the inverse of the cell size by convention. If we don't do this, the resulting grid will be transposed.

Note that we define filenames for slope and aspect grids, which are two intermediate products that are combined to create the final product. These intermediate grids are output as well. They can also serve as inputs to other types of products.

This script uses a three-by-three windowing method to scan the image and smooth out the center value in these mini grids to process the image efficiently and within the memory constraints of your computer. However, because we are using NumPy, we can process the entire array at once via matrices, as opposed to a lengthy series of nested loops. This technique is based on the excellent work of a developer named Michal Migurski, who implemented the clever NumPy version of Matthew Perry's C++ implementation, which served as the basis for the DEM tools in the GDAL suite.

After **slope** and **aspect** are calculated, they are used to output the shaded relief. The slope is the steepness of a hill or mountain, while the aspect is the direction the grid cell faces that is specified as a degree between 0 and 360. Finally, everything is saved to the disk from NumPy. In the savetxt() method, we specify a four-integer format string as the peak elevations are several thousand meters:

```
from linecache import getline
import numpy as np

# File name of ASCII digital elevation model
source = "dem.asc"
# File name of the slope grid
slopegrid = "slope.asc"
# File name of the aspect grid
aspectgrid = "aspect.asc"
# Output file name for shaded relief
shadegrid = "relief.asc"

# Shaded elevation parameters
# Sun direction
azimuth = 315.0
# Sun angle
```

```
altitude = 45.0
# Elevation exagerationexaggeration
z = 1.0
# Resolution
scale = 1.0
# No data value for output
NODATA = -9999

# Needed for numpy conversions
deg2rad = 3.141592653589793 / 180.0
rad2deg = 180.0 / 3.141592653589793

# Parse the header using a loop and
# the built-in linecache module
hdr = [getline(source, i) for i in range(1, 7)]
values = [float(h.split(" ")[-1].strip()) for h in hdr]
cols, rows, lx, ly, cell, nd = values
xres = cell
yres = cell * -1

# Load the dem into a numpy array
arr = np.loadtxt(source, skiprows=6)

# Exclude 2 pixels around the edges which are usually NODATA.
# Also set up structure for 3x3 windows to process the slope
# throughout the grid
window = []
for row in range(3):
    for col in range(3):
        window.append(arr[row:(row + arr.shape[0] - 2),
                      col:(col + arr.shape[1] - 2)])

# Process each 3x3 window in both the x and y directions
x = ((z * window[0] + z * window[3] + z * window[3] + z *
  window[6]) -
    (z * window[2] + z * window[5] + z * window[5] + z *
    window[8])) / \
    (8.0 * xres * scale)

y = ((z * window[6] + z * window[7] + z * window[7] + z *
  window[8]) -
    (z * window[0] + z * window[1] + z * window[1] + z *
    window[2])) / \
```

```
        (8.0 * yres * scale)

# Calculate slope
slope = 90.0 - np.arctan(np.sqrt(x * x + y * y)) * rad2deg

# Calculate aspect
aspect = np.arctan2(x, y)

# Calculate the shaded relief
shaded = np.sin(altitude * deg2rad) * np.sin(slope * deg2rad) + \
    np.cos(altitude * deg2rad) * np.cos(slope * deg2rad) * \
    np.cos((azimuth - 90.0) * deg2rad - aspect)
# Scale values from 0-1 to 0-255
shaded = shaded * 255

# Rebuild the new header
header = "ncols          {}\n".format(shaded.shape[1])
header += "nrows          {}\n".format(shaded.shape[0])
header += "xllcorner      {}\n".format(lx + (cell * (cols -
    shaded.shape[1])))
header += "yllcorner      {}\n".format(ly + (cell * (rows -
    shaded.shape[0])))
header += "cellsize       {}\n".format(cell)
header += "NODATA_value       {}\n".format(NODATA)

# Set no-data values
for pane in window:
    slope[pane == nd] = NODATA
    aspect[pane == nd] = NODATA
    shaded[pane == nd] = NODATA

# Open the output file, add the header, save the slope grid
with open(slopegrid, "wb") as f:
    f.write(bytes(header, "UTF-8")
    np.savetxt(f, slope, fmt="%4i")

# Open the output file, add the header, save the aspectgrid
with open(aspectgrid, "wb") as f:
    f.write(bytes(header, "UTF-8")
    np.savetxt(f, aspect, fmt="%4i")

# Open the output file, add the header, save the relief grid
with open(shadegrid, "wb") as f:
    f.write(bytes(header, 'UTF-8'))
    np.savetxt(f, shaded, fmt="%4i")
```

If we load the output shaded relief grid to QGIS and specify the styling to stretch the image to the minimum and maximum, we can see the following image. If QGIS asks you for a projection, the data is `EPSG:3157`. You can also open the image in the FWTools OpenEV application discussed in the *Installing GDAL* section in *Chapter 4, Geospatial Python Toolbox*, which will automatically stretch the image for optimal viewing:

As you can see, the preceding image is much easier to comprehend than the original pseudo-color representation that we examined originally. Next, let's look at the slope raster used to create the shaded relief:

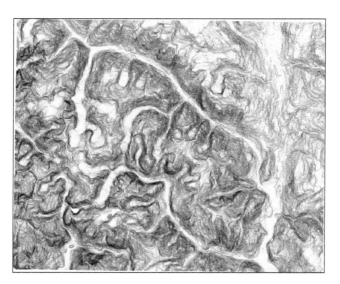

The slope shows the gradual decline in elevation from the high points to low points in all the directions of the dataset. Slope is an especially useful input for many types of hydrology models.

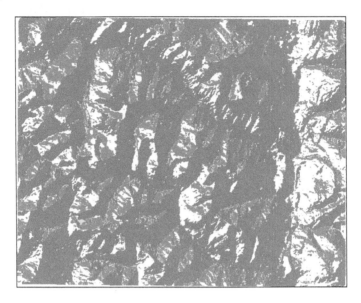

The aspect shows the maximum rate of downslope change from one cell to its neighbors. If you compare the aspect image to the shaded relief image, you can see that the red and gray values of the aspect image correspond to shadows in the shaded relief. So, the slope is primarily responsible for turning the DEM into a terrain relief while the aspect is responsible for the shading.

Creating elevation contours

Now, let's look at another way to visualize the elevation better using contours. A contour is an isoline along the same elevation in a dataset. Contours are usually stepped at intervals to create an intuitive way to represent elevation data, both visually and numerically, using a resource-efficient vector dataset.

The input to generate contours is our DEM and the output is a shapefile. The algorithm to generate contours is fairly complex and very difficult to implement using NumPy's linear algebra. So, our solution in this case is to fall back on the GDAL library, which has a contouring method available through the Python API. In fact, the majority of this script is just setting up the OGR library code that is needed to output a shapefile. The actual contouring is a single method call named `gdal.ContourGenerate()`. Just before this call, there are comments defining the method's arguments. The most important ones are as follows:

- `contourInterval`: It is the distance in the dataset units between contours
- `contourBase`: It is the starting elevation for the contouring
- `fixedLevelCount`: It specifies a fixed number of contours as opposed to distance
- `idField`: It is a name for a required shapefile dbf field, usually just called ID
- `elevField`: It is a name for a required shapefile dbf field for the elevation value that is useful for the labeling in maps

You should have GDAL and OGR installed from the *Installing GDAL* section in *Chapter 4*, *Geospatial Python Toolbox*. In the following code, we will define the input DEM filename, output shapefile name, create the shapefile data source with OGR, get the OGR layer, open the DEM, and generate contours on the OGR layer:

```
import gdal
import ogr

# Elevation DEM
source = "dem.asc"
# Output shapefile
target = "contour"

ogr_driver = ogr.GetDriverByName("ESRI Shapefile")
ogr_ds = ogr_driver.CreateDataSource(target + ".shp")
ogr_lyr = ogr_ds.CreateLayer(target,
# wkbLineString25D is the type code for geometry with a z
# elevation value.
geom_type=ogr.wkbLineString25D)
field_defn = ogr.FieldDefn("ID" ogr.OFTInteger)
ogr_lyr.CreateField(field_defn)
field_defn = ogr.FieldDefn("ELEV" ogr.OFTReal)
ogr_lyr.CreateField(field_defn)

# gdal.ContourGenerate() arguments
# Band srcBand,
```

```
# double contourInterval,
# double contourBase,
# double[] fixedLevelCount,
# int useNoData,
# double noDataValue,
# Layer dstLayer,
# int idField,
# int elevField
ds = gdal.Open(source)
# EPGS:3157
gdal.ContourGenerate(ds.GetRasterBand(1), 400, 10, [], 0, 0, ogr_lyr,
0, 1))
ogr_ds = None
```

Now, let's draw the contour shapefile that we just created using PNGCanvas,
introduced in the *PNGCanvas* section of *Chapter 4, Geospatial Python Toolbox*:

```
import shapefile
import pngcanvas
# Open the contours
r = shapefile.Reader("contour.shp")
# Setup the world to pixels conversion
xdist = r.bbox[2] - r.bbox[0]
ydist = r.bbox[3] - r.bbox[1]
iwidth = 800
iheight = 600
xratio = iwidth/xdist
yratio = iheight/ydist
contours = []
# Loop through all shapes
for shape in r.shapes():
    # Loop through all parts
    for i in range(len(shape.parts)):
        pixels = []
        pt = None
        if i < len(shape.parts) - 1:
            pt = shape.points[shape.parts[i]:shape.parts[i+1]]
        else:
            pt = shape.points[shape.parts[i]:]
        for x, y in pt:
            px = int(iwidth - ((r.bbox[2] - x) * xratio))
            py = int((r.bbox[3] - y) * yratio)
            pixels.append([px, py])
        contours.append(pixels)
```

```
# Set up the output canvas
canvas = pngcanvas.PNGCanvas(iwidth, iheight)
# PNGCanvas accepts rgba byte arrays for colors
red = [0xff, 0, 0, 0xff]
canvas.color = red
# Loop through the polygons and draw them
for c in contours:
    canvas.polyline(c)
# Save the image
with open("contours.png", "wb") as f:
    f.write(canvas.dump())
```

We will end up with the following image:

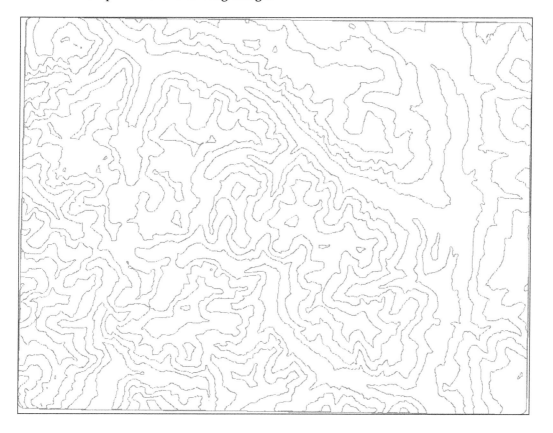

If we bring our shaded relief ASCIIGRID and the shapefile into a GIS, such as QGIS, we can create a simple topographic map as follows. You can use the elevation (that is *ELEV*) dbf field that you specified in the script to label the contour lines with the elevation:

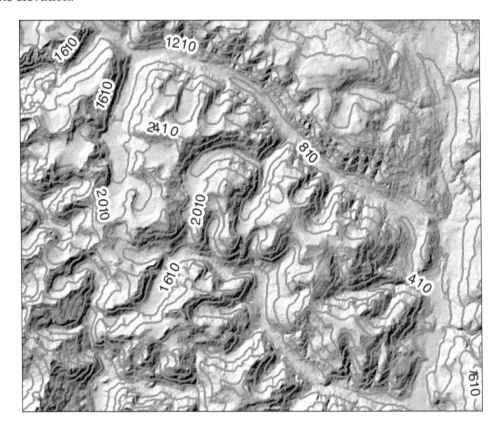

The techniques in these NumPy grid examples provide the building blocks for all kinds of elevation products. The USGS has an excellent web page with sample elevation-based data layers, including the examples that we created as well as some more advanced types:

http://edna.usgs.gov/datalayers.asp

Next, we'll work with one of the most complex elevation data types: **LIDAR** data.

Working with LIDAR

LIDAR stands for **Light Detection and Ranging**. It is similar to radar-based images but uses finite laser beams, which hit the ground hundreds of thousands of times per second to collect a huge amount of very fine (x, y, z) locations as well as time and intensity. The intensity value is what really separates LIDAR from other data types. For example, the asphalt rooftop of a building may be of the same elevation as the top of a nearby tree, but the intensities will be different. Just like the remote sensing radiance values in a multispectral satellite image allow us to build classification libraries, the intensity values of LIDAR data allow us to classify and colorize LIDAR data.

The high volume and precision of LIDAR actually makes it difficult to use. A LIDAR dataset is referred to as a **point cloud** because the shape of the dataset is usually irregular as the data is three-dimensional with outlying points. There are not many software packages that effectively visualize point clouds. Furthermore, an irregular-shaped collection of finite points is just hard to interact with, even when using appropriate software.

For these reasons, one of the most common operations on LIDAR data is to project the data and resample it to a regular grid. We'll do exactly this using a small LIDAR dataset. This dataset is approximately 7 MB uncompressed and contains over 600,000 points. The data captures some easily identifiable features, such as buildings, trees, and cars in parking lots. You can download the zipped dataset at `http://git.io/vOERW`.

The file format is a very common binary format specific to LIDAR called LAS, which is short for **laser**. Unzip this file to your working directory. In order to read this format, we'll use a pure Python library called `laspy`. You can install the version for Python 3 using the following command:

```
pip install http://git.io/vOER9
```

Creating a grid from LIDAR

With `laspy` installed, we are ready to create a grid from LIDAR. This script is fairly straightforward. We loop through the (x,y) point locations in the LIDAR data and project them to our grid with a cell size of one meter. Due to the precision of the LIDAR data, we'll end up with multiple points in a single cell. We average these points to create a common elevation value. Another issue that we have to deal with is data loss. Whenever you resample the data, you lose information. In this case, we'll end up with NODATA holes in the middle of the raster. To deal with this issue, we fill these holes with average values from the surrounding cells, which is a form of interpolation.

We only need two modules, both available on PyPI, as shown in the following code:

```
from laspy.file import File
import numpy as np

# Source LAS file
source = "lidar.las"

# Output ASCII DEM file
target = "lidar.asc"

# Grid cell size (data units)
cell = 1.0

# No data value for output DEM
NODATA = 0

# Open LIDAR LAS file
las = File(source, mode="r")

# xyz min and max
min = las.header.min
max = las.header.max

# Get the x axis distance im meters
xdist = max[0] - min[0]

# Get the y axis distance in meters
ydist = max[1] - min[1]

# Number of columns for our grid
cols = int(xdist) / cell

# Number of rows for our grid
rows = int(ydist) / cell

cols += 1
rows += 1

# Track how many elevation
# values we aggregate
count = np.zeros((rows, cols)).astype(np.float32)
# Aggregate elevation values
```

```python
zsum = np.zeros((rows, cols)).astype(np.float32)

# Y resolution is negative
ycell = -1 * cell

# Project x, y values to grid
projx = (las.x - min[0]) / cell
projy = (las.y - min[1]) / ycell
# Cast to integers and clip for use as index
ix = projx.astype(np.int32)
iy = projy.astype(np.int32)

# Loop through x, y, z arrays, add to grid shape,
# and aggregate values for averaging
for x, y, z in np.nditer([ix, iy, las.z]):
    count[y, x] += 1
    zsum[y, x] += z

# Change 0 values to 1 to avoid numpy warnings,
# and NaN values in array
nonzero = np.where(count > 0, count, 1)
# Average our z values
zavg = zsum / nonzero

# Interpolate 0 values in array to avoid any
# holes in the grid
mean = np.ones((rows, cols)) * np.mean(zavg)
left = np.roll(zavg, -1, 1)
lavg = np.where(left > 0, left, mean)
right = np.roll(zavg, 1, 1)
ravg = np.where(right > 0, right, mean)
interpolate = (lavg + ravg) / 2
fill = np.where(zavg > 0, zavg, interpolate)

# Create our ASCII DEM header
header = "ncols         {}\n".format(fill.shape[1])
header += "nrows         {}\n".format(fill.shape[0])
header += "xllcorner     {}\n".format(min[0])
header += "yllcorner     {}\n".format(min[1])
header += "cellsize      {}\n".format(cell)
header += "NODATA_value      {}\n".format(NODATA)

# Open the output file, add the header, save the array
with open(target, "wb") as f:
```

```
f.write(bytes(header, 'UTF-8'))
# The fmt string ensures we output floats
# that have at least one number but only
# two decimal places
np.savetxt(f, fill, fmt="%1.2f")
```

The result of our script is an ASCIIGRID, which looks like the following image when viewed in OpenEV. Higher elevations are lighter while lower elevations are darker. Even in this form, you can see buildings, trees, and cars:

If we assigned a heat map color ramp, the colors give you a sharper sense of the elevation differences:

So, what happens if we run this output DEM through our shaded relief script from earlier? There's a big difference between straight-sided buildings and sloping mountains. If you change the input and output names in the shaded relief script to process the LIDAR DEM, we get the following slope result:

The gently rolling slope of the mountainous terrain is reduced to outlines of major features in the image. In the aspect image, the changes are so sharp and over such short distances that the output image is very chaotic to view, as shown in the following screenshot:

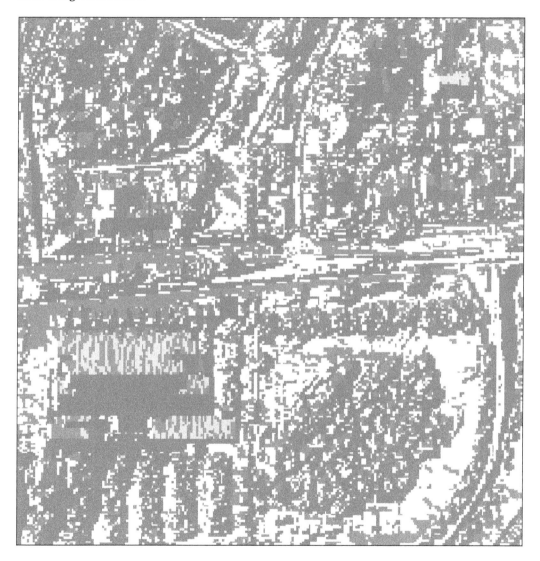

Despite the difference in these images and the coarser but smoother mountain versions, we still get a very nice shaded relief, which visually resembles a black and white photograph:

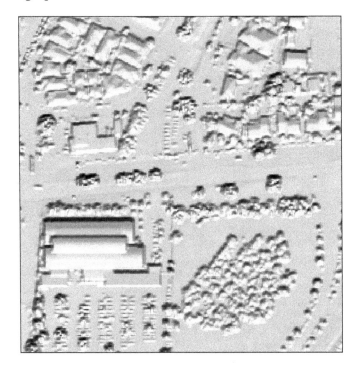

Using PIL to visualize LIDAR

The previous DEM images in this chapter were visualized using QGIS and OpenEV. We can also create output images in Python by introducing some new functions of the **Python Imaging Library (PIL)**, which we didn't use in the previous chapters. In this example, we'll use the PIL.ImageOps module, which has functions for histogram equalization and automatic contrast enhancement. We'll use PIL's fromarray() method to import the data from NumPy. Let's see how close we can get to the output of the desktop GIS programs pictured in this chapter with the help of the following code:

```
import numpy as np

try:
    import Image
    import ImageOps
except:
```

```
    from PIL import Image, ImageOps

# Source gridded LIDAR DEM file
source = "lidar.asc"

# Output image file
target = "lidar.bmp"

# Load the ASCII DEM into a numpy array
arr = np.loadtxt(source, skiprows=6)

# Convert array to numpy image
im = Image.fromarray(arr).convert("RGB")

# Enhance the image:
# equalize and increase contrast
im = ImageOps.equalize(im)
im = ImageOps.autocontrast(im)

# Save the image
im.save(target)
```

As you can see in the following screenshot, the enhanced shaded relief has a sharper relief than the previous version:

Now let's colorize our shaded relief. We'll use the built-in Python `colorsys` module for color space conversion. Normally, we specify colors as RGB values. However, to create a color ramp for a heat map scheme, we'll use **HSV** values, which stand for **Hue**, **Saturation**, and **Value**, to generate our colors. The advantage of HSV is that you can tweak the H value as a degree between zero and 360 on a color wheel. Using a single value for hue allows you to use a linear ramping equation, which is much easier than trying to deal with combinations of three separate RGB values. The following image taken from the online magazine, *Qt Quarterly*, illustrates the HSV color model:

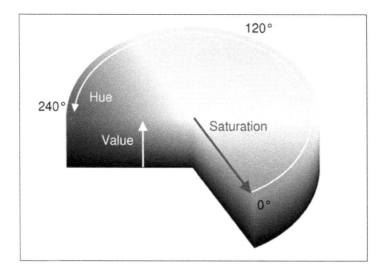

The `colorsys` module lets you switch back and forth between the HSV and RGB values. The module returns percentages for RGB values, which must then be mapped to the 0-255 scale for each color.

In the following code, we'll convert the ASCII DEM to a PIL image, build our color palette, apply the color palette to the grayscale image, and save the image:

```
import numpy as np
try:
    import Image
    import ImageOps
except:
    from PIL import Image, ImageOps
import colorsys

# Source LIDAR DEM file
```

```
source = "lidar.asc"

# Output image file
target = "lidar.bmp"

# Load the ASCII DEM into a numpy array
arr = np.loadtxt(source, skiprows=6)

# Convert the numpy array to a PIL image.
# Use black and white mode so we can stack
# three bands for the color image.
im = Image.fromarray(arr).convert('L')

# Enhance the image
im = ImageOps.equalize(im)
im = ImageOps.autocontrast(im)

# Begin building our color ramp
palette = []

# Hue, Saturation, Value
# color space starting with yellow.
h = .67
s = 1
v = 1

# We'll step through colors from:
# blue-green-yellow-orange-red.
# Blue=low elevation, Red=high-elevation
step = h / 256.0

# Build the palette
for i in range(256):
    rp, gp, bp = colorsys.hsv_to_rgb(h, s, v)
    r = int(rp * 255)
    g = int(gp * 255)
    b = int(bp * 255)
    palette.extend([r, g, b])
    h -= step

# Apply the palette to the image
im.putpalette(palette)

# Save the image
im.save(target)
```

The code produces the following image with higher elevations in warmer colors and lower elevations in cooler colors:

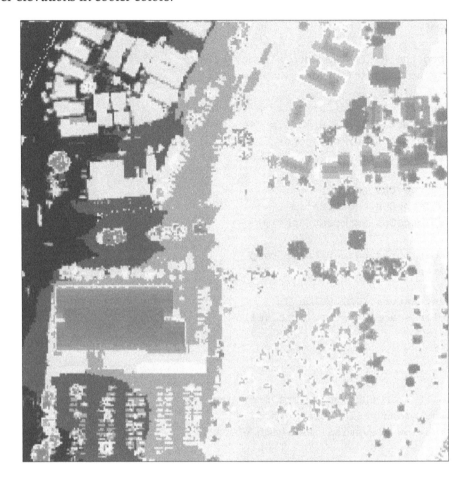

In this image, we actually get more variation than the default QGIS version. We could potentially improve this image with a smoothing algorithm that would blend the colors where they meet and soften the image visually. As you can see, we have the full range of our color ramp expressed from cool to warm colors as the elevation change increases.

Creating a triangulated irregular network

The following example is our most sophisticated example yet. A **triangulated irregular network (TIN)** is a vector representation of a point dataset in a vector surface of points connected as triangles. An algorithm determines which points are absolutely necessary to accurately represent the terrain as opposed to a raster, which stores a fixed number of cells over a given area and may repeat elevation values in adjacent cells that could be more efficiently stored as a polygon. A TIN can also be resampled more efficiently on the fly than a raster, which requires less computer memory and processing power when using TIN in a GIS. The most common type of TIN is based on **Delaunay triangulation**, which includes all the points without redundant triangles.

The Delaunay triangulation is very complex. We'll use a pure Python library written by Bill Simons as a part of Steve Fortune's Delaunay triangulation algorithm called `voronoi.py` to calculate the triangles in our LIDAR data. You can download the script to your working directory or site-packages directory from `http://git.io/vOEuJ`.

This script reads the LAS file, generates the triangles, then loops through them, and writes out a shapefile. For this example, we'll use a clipped version of our LIDAR data to reduce the area to process. If we run our entire dataset of 600,000 plus points, the script will run for hours and generate over half a million triangles. You can download the clipped LIDAR dataset as a ZIP file from the following URL:

`http://git.io/vOE62`

We have several status messages that print while the script runs because of the time-intensive nature of the following example, which can take several minutes to be complete. We'll be storing the triangles as `PolygonZ` types, which allow the vertices to have a z elevation value. Unzip the LAS file and run the following code to generate a shapefile called `mesh.shp`:

```
import pickle
import os
import time
import math
import numpy as np
import shapefile

# laspy for Python 3: pip install http://git.io/vOER9
from laspy.file import File

# voronoi.py for Python 3: pip install http://git.io/vOEuJ
```

```
import voronoi

# Source LAS file
source = "clippedLAS.las"

# Output shapefile
target = "mesh"

# Triangles pickle archive
archive = "triangles.p"

# Pyshp archive
pyshp = "mesh_pyshp.p"


class Point:
    """Point class required by the voronoi module"""
    def __init__(self, x, y):
        self.px = x
        self.py = y

    def x(self):
        return self.px

    def y(self):
        return self.py

# The triangle array holds tuples
# 3 point indicesused to retrieve the points.
# Load it from a pickle
# file or use the voronoi module
# to create the triangles.
triangles = None

if os.path.exists(archive):
    print("Loading triangle archive...")
    f = open(archive, "rb")
    triangles = pickle.load(f)
    f.close()
    # Open LIDAR LAS file
    las = File(source, mode="r")
else:
    # Open LIDAR LAS file
```

```
las = File(source, mode="r")
points = []
print("Assembling points...")
# Pull points from LAS file
for x, y in np.nditer((las.x, las.y)):
    points.append(Point(x, y))
print("Composing triangles...")
# Delaunay Triangulation
triangles = voronoi.computeDelaunayTriangulation(points)
# Save the triangles to save time if we write more than
# one shapefile.
f = open(archive, "wb")
pickle.dump(triangles, f, protocol=2)
f.close()

print("Creating shapefile...")
w = None
if os.path.exists(pyshp):
    f = open(pyshp, "rb")
    w = pickle.load(f)
    f.close()
else:
    # PolygonZ shapefile (x, y, z, m)
    w = shapefile.Writer(shapefile.POLYGONZ)
    w.field("X1", "C", "40")
    w.field("X2", "C", "40")
    w.field("X3", "C", "40")
    w.field("Y1", "C", "40")
    w.field("Y2", "C", "40")
    w.field("Y3", "C", "40")
    w.field("Z1", "C", "40")
    w.field("Z2", "C", "40")
    w.field("Z3", "C", "40")
    tris = len(triangles)
    # Loop through shapes and
    # track progress every 10 percent
    last_percent = 0
    for i in range(tris):
        t = triangles[i]
        percent = int((i/(tris*1.0))*100.0)
        if percent % 10.0 == 0 and percent > last_percent:
            last_percent = percent
```

```
        print("{} % done - Shape {}/{} at {}".format(percent,
            i, tris, time.asctime()))
    part = []
    x1 = las.x[t[0]]
    y1 = las.y[t[0]]
    z1 = las.z[t[0]]
    x2 = las.x[t[1]]
    y2 = las.y[t[1]]
    z2 = las.z[t[1]]
    x3 = las.x[t[2]]
    y3 = las.y[t[2]]
    z3 = las.z[t[2]]
    # Check segments for large triangles
    # along the convex hull which is acommon
    # artificat in Delaunay triangulation
    max = 3
    if math.sqrt((x2-x1)**2+(y2-y1)**2) > max:
        continue
    if math.sqrt((x3-x2)**2+(y3-y2)**2) > max:
        continue
    if math.sqrt((x3-x1)**2+(y3-y1)**2) > max:
        continue
    part.append([x1, y1, z1, 0])
    part.append([x2, y2, z2, 0])
    part.append([x3, y3, z3, 0])
    w.poly(parts=[part])
    w.record(x1, x2, x3, y1, y2, y3, z1, z2, z3)
print("Saving shapefile...")
# Pickle the Writer in case something
# goes wrong. Be sure to delete this
# file to recreate theshapefile.
f = open(pyshp, "wb")
pickle.dump(w, f, protocol=2)
f.close()
w.save(target)
print("Done.")
```

The following image shows a zoomed-in version of the TIN over the colorized LIDAR data:

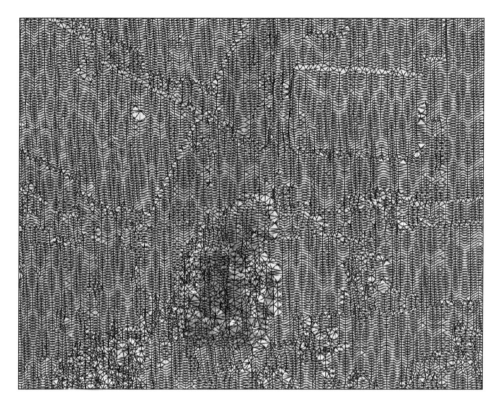

Summary

Elevation data can often provide a complete dataset for analysis and derivative products without any other data. In this chapter, you learned to read and write ASCII Grids using only NumPy. You also learned to create shaded reliefs, slope grids, and aspect grids. We created elevation contours using a little-known feature of the GDAL library available for Python. We transformed LIDAR data into an easy-to-manipulate ASCII Grid. We experimented with different ways to visualize the LIDAR data with the PIL. Finally, we created a 3D surface or TIN by turning a LIDAR point cloud into a 3D shapefile of polygons.

In the next chapter, we'll combine the building blocks from the previous three chapters to perform some advanced modeling and actually create some information products.

8
Advanced Geospatial Python Modeling

In this chapter, we'll build on the data processing concepts that you've learned up to this point to create some full-scale information products. We will introduce some important geospatial algorithms commonly used in agriculture, emergency management, logistics, and other industries.

We'll create the following products in this chapter:

- A crop health map
- A flood inundation model
- A colorized hillshade
- A terrain routing map
- A street routing map
- A shapefile with links to geolocated photos

While these products are task specific, the algorithms used to create them are widely applied in geospatial analysis. The examples in this chapter are longer and more complex than in the previous chapters. For that reason, there are far more code comments to make the programs easier to follow. We will also use more functions in these examples. In the previous chapters, functions were mostly avoided for clarity. But these examples are sufficiently complex, and certain functions serve to make the code easier to read.

Creating a Normalized Difference Vegetative Index

Our first example will be a **Normalized Difference Vegetation Index** or **NDVI**. NDVIs are used to show the relative health of plants in an area of interest. An NDVI algorithm uses satellite or aerial imagery to show relative health by highlighting chlorophyll density in plants. NDVIs use only the red and near-infrared bands. Take a look at the following formula for this:

NDVI = (Infrared – Red) / (Infrared + Red)

The goal of this analysis is to begin with a multispectral image containing those two bands and end up with a pseudo-color image using seven classes that color the healthier plants darker green, less healthy plants lighter green, and bare soil brown.

Because the health index is relative, it is important to localize the area of interest. You could perform a relative index for the entire globe, but vast areas like the Sahara Desert on the low-vegetation extreme and densely forested areas like the Amazon jungle skew the results for vegetation in the middle range. However, that being said, climate scientists do routinely create global NDVIs to study worldwide trends. The more common application, though, is for managed areas such as a forest or a farm field, as shown in this example.

We will begin with analysis of a single farm field in the Mississippi Delta. To do so, we'll start with a multispectral image of a fairly large area and use a shapefile to isolate a single field. The image in the following screenshot is our broad area with the field of interest highlighted in yellow:

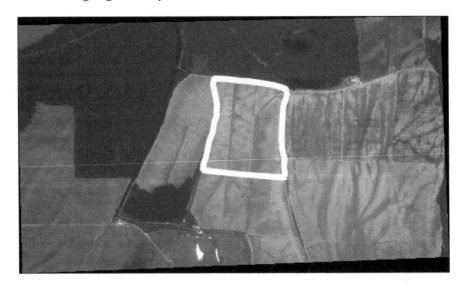

You can download this image and the shapefile for the farm field as a ZIP file from `http://git.io/v3fS9`.

For this example, we'll use GDAL, OGR, `gdal_array`/NumPy, and PIL to clip and process the data. In the other examples in this chapter, we'll use simple ASCII Grids and NumPy only; as we'll be using ASCII elevation grids, GDAL isn't required. In all the examples, the scripts use the following convention:

1. Import libraries.
2. Define functions.
3. Define global variables such as filenames.
4. Execute the analysis.
5. Save the output.

Our approach to the crop health example is split into two scripts. The first script creates the index image, which is a grayscale image. The second script classifies the index and outputs a colored image.

In this first script, we will execute the following steps:

1. Read the red band.
2. Read the infrared band.
3. Read the field boundary shapefile.
4. Rasterize the shapefile to an image.
5. Convert the shapefile image to a NumPy array.
6. Use the NumPy array to clip the red band to the field.
7. Do the same for the infrared band.
8. Use the band arrays to execute the NDVI algorithm in NumPy.
9. Save the resulting indexing algorithm to a GeoTIFF using `gdal_array`.

We'll discuss this script in sections to make it easier to follow. The code comments will also tell you what is going on at each step of the way.

Setting up the framework

This section imports the modules we need and sets up the functions that we'll use for the preceding steps 1 to 5. The `imageToArray()` function converts a PIL image to a NumPy array and is dependent on the `gdal_array` and `PIL` modules. The `world2Pixel()` function converts geospatial coordinates to the pixel coordinates of our target image. This function uses the georeferencing information presented by the `gdal` module. The `copy_geo()` function copies the georeferencing information from our source image to our target array; however, it accounts for the offset created when we clip the image. These functions are fairly generic and can serve a role in a variety of different remote sensing processes beyond the example given here:

```python
import gdal
from osgeo import gdal
from osgeo import gdal_array
from osgeo import ogr
try:
    import Image
    import ImageDraw
except:
    from PIL import Image, ImageDraw


def imageToArray(i):
    """
    Converts a Python Imaging Library
    array to a gdal_array image.
    """
    a = gdal_array.numpy.fromstring(i.tostring(), 'b')
    a.shape = i.im.size[1], i.im.size[0]
    return a


def world2Pixel(geoMatrix, x, y):
    """
    Uses a gdal geomatrix (gdal.GetGeoTransform())
    to calculate the pixel location of a
    geospatial coordinate
    """
    ulX = geoMatrix[0]
    ulY = geoMatrix[3]
    xDist = geoMatrix[1]
    yDist = geoMatrix[5]
    rtnX = geoMatrix[2]
    rtnY = geoMatrix[4]
```

```
    pixel = int((x - ulX) / xDist)
    line = int((ulY - y) / abs(yDist))
    return (pixel, line)
def copy_geo(array, prototype=None, xoffset=0, yoffset=0):
    """Copy geotransfrom from prototype dataset to array but account
    for x, y offset of clipped array."""
    ds = gdal.Open(gdal_array.GetArrayFilename(array))
    prototype = gdal.Open(prototype)
    gdal_array.CopyDatasetInfo(prototype, ds,
                               xoff=xoffset, yoff=yoffset)
    return ds
```

Loading the data

In this section, we load the source image of a farm field using `gdal_array` that takes it straight into a NumPy array. We also define the name of our output image, which will be `ndvi.tif`. One interesting piece of this section is that we load the source image a second time using the `gdal` module as opposed to `gdal_array`. This second call is to capture the georeferencing data for the image that is available through `gdal` and not `gdal_array`. Fortunately, `gdal` only loads raster data on demand, so this approach avoids loading the complete dataset into memory twice. Once we have the data as a multi-dimensional NumPy array, we split out the red and infrared bands, as they will both be used in the NDVI equation, as shown here:

```
# Multispectral image used
# to create the NDVI. Must
# have red and infrared
# bands
source = "farm.tif"

# Output geotiff file name
target = "ndvi.tif"

# Load the source data as a gdal_array array
srcArray = gdal_array.LoadFile(source)

# Also load as a gdal image to
# get geotransform info
srcImage = gdal.Open(source)
geoTrans = srcImage.GetGeoTransform()

# Red and infrared (or near infrared) bands
r = srcArray[1]
ir = srcArray[2]
```

Rasterizing the shapefile

This section begins the process of clipping. However, the first step is to rasterize the shapefile outlining the boundry of the specific area that we are going to analyze in the larger `field.tif` satellite image that was previously loaded; in other words, you need to convert it from vector data to raster data. But, we also want to fill in the polygon when we convert it, so it can be used as an image mask. The pixels in the mask will be correlated to the pixels in the red and infrared arrays. Any pixels outside the mask will be turned to NODATA pixels, so they are not processed as part of the NDVI. To make this correlation, we'll need the solid polygon to be a NumPy array just like the raster bands. This approach will make sure our NDVI calculation will be limited to the farm field.

The easiest way to convert the shapefile polygon into a filled polygon as a NumPy array is to plot it as a polygon in a PIL image, fill that polygon in, and then convert it to a NumPy array using existing methods in PIL and NumPy that allow that conversion. In this example, we use the ogr module to read the shapefile because we already have GDAL available. But, we could have also used PyShp to read the shapefile just as easily as the ogr module. If our farm field image was available as an ASCII Grid, we could have avoided using the gdal, gdal_array, and ogr modules altogether, as shown here:

```
# Clip a field out of the bands using a
# field boundary shapefile

# Create an OGR layer from a Field boundary shapefile
field = ogr.Open("field.shp")
# Must define a "layer" to keep OGR happy
lyr = field.GetLayer("field")
# Only one polygon in this shapefile
poly = lyr.GetNextFeature()

# Convert the layer extent to image pixel coordinates
minX, maxX, minY, maxY = lyr.GetExtent()
ulX, ulY = world2Pixel(geoTrans, minX, maxY)
lrX, lrY = world2Pixel(geoTrans, maxX, minY)

# Calculate the pixel size of the new image
pxWidth = int(lrX - ulX)
pxHeight = int(lrY - ulY)

# Create a blank image of the correct size
# that will serve as our mask
clipped = gdal_array.numpy.zeros((3, pxHeight, pxWidth),
```

```
                              gdal_array.numpy.uint8)

# Clip red and infrared to new bounds.
rClip = r[ulY:lrY, ulX:lrX]
irClip = ir[ulY:lrY, ulX:lrX]

# Create a new geomatrix for the image
geoTrans = list(geoTrans)
geoTrans[0] = minX
geoTrans[3] = maxY

# Map points to pixels for drawing
# the field boundary on a blank
# 8-bit, black and white, mask image.
points = []
pixels = []
# Grab the polygon geometry
geom = poly.GetGeometryRef()
pts = geom.GetGeometryRef(0)

# Loop through geometry and turn
# the points into an easy-to-manage
# Python list
for p in range(pts.GetPointCount()):
    points.append((pts.GetX(p), pts.GetY(p)))

# Loop through the points and map to pixels.
# Append the pixels to a pixel list
for p in points:
    pixels.append(world2Pixel(geoTrans, p[0], p[1]))

# Create the raster polygon image as a black and white 'L' mode
# and filled as white.  White=1
rasterPoly = Image.new("L", (pxWidth, pxHeight), 1)

# Create a PIL drawing object
rasterize = ImageDraw.Draw(rasterPoly)

# Dump the pixels to the image
# as a polygon. Black=0
rasterize.polygon(pixels, 0)

# Hand the image back to gdal/gdal_array
# so we can use it as an array mask
mask = imageToArray(rasterPoly)
```

Clipping the bands

Now that we have our image mask, we can clip the red and infrared bands to the boundary of the mask. For this process, we use NumPy's `choose()` method that correlates the mask cell to the raster band cell and returns that value or returns 0. The result is a new array that is clipped to the mask but with the correlated values from the raster band, as shown here:

```
# Clip the red band using the mask
rClip = gdal_array.numpy.choose(mask,
                        (rClip, 0)).astype(gdal_array.numpy.uint8)

# Clip the infrared band using the mask
irClip = gdal_array.numpy.choose(mask,(irClip,
    0)).astype(gdal_array.numpy.uint8)
```

Using the NDVI formula

The final process for creating the NDVI is to execute the equation that is *infrared – red / infrared + red*. The first step we perform silences any *not-a-number* also known as NaN values in NumPy that might occur during division. And before we save the output, we'll convert any NaN values to 0. We'll save the output as `ndvi.tif`, which will be the input for the next script to classify and colorize the NDVI, as shown in the following lines of code:

```
# We don't care about numpy warnings
# due to NaN values from clipping
gdal_array.numpy.seterr(all="ignore")

# NDVI equation: (infrared - red) / (infrared + red)
# *1.0 converts values to floats,
# +1.0 prevents ZeroDivisionErrors
ndvi = 1.0 * ((irClip - rClip) / (irClip + rClip + 1.0))

# Convert any NaN values to 0 from the final product
ndvi = gdal_array.numpy.nan_to_num(ndvi)

# Save the ndvi as a  GeoTIFF and copy/adjust
# the georeferencing info
gtiff = gdal.GetDriverByName( 'GTiff' )
gtiff.CreateCopy(target, copy_geo(ndvi, prototype=source,
xoffset=ulX, yoffset=ulY))
gtiff = None
```

Take a look at the following image which is the output of this example. You need to view it in a geospatial viewer such as QGIS or OpenEV. The image won't open in most image editors. You can see that the lighter the shade of gray, the healthier the plant is within that field:

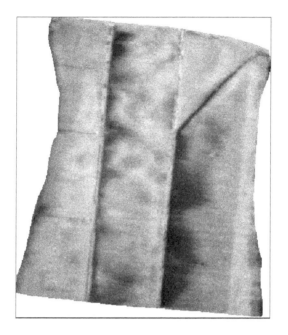

Classifying the NDVI

We now have a valid index, but it is not easy to understand because it is a grayscale image. If we color the image in an intuitive way, then even a child can identify the healthier plants. The following example reads in this grayscale index and classifies it from brown to dark green using seven classes. The classification and image processing routines such as the histogram and stretching functions are almost identical to what we used in the *Creating histograms* section, in *Chapter 6, Python and Remote Sensing*, but this time, we are applying them in a much more specific way. The output of this example will be another GeoTIFF, but this time it will be a colorful RGB image.

Additional functions

We won't need any of the functions from our previous NDVI script, but we do need to add a function for creating and stretching a histogram. Both of these functions work with NumPy arrays. We'll also shorten the reference to `gdal_array` in this script to `gd` because it is a long name and we need it throughout this script, as shown in the following lines of code:

```
import gdal_array as gd
import operator
from functools import reduce


def histogram(a, bins=list(range(256))):
    """
    Histogram function for multi-dimensional array.
    a = array
    bins = range of numbers to match
    """
# Flatten, sort, then split our arrays for the histogram.
    fa = a.flat
    n = gd.numpy.searchsorted(gd.numpy.sort(fa), bins)
    n = gd.numpy.concatenate([n, [len(fa)]])
    hist = n[1:]-n[:-1]
    return hist


def stretch(a):
    """
    Performs a histogram stretch on a gdal_array array image.
    """
    hist = histogram(a)
    lut = []
    for b in range(0, len(hist), 256):
        # step size—create equal interval bins.
        step = reduce(operator.add, hist[b:b+256]) / 255
        # create equalization lookup table
        n = 0
        for i in range(256):
            lut.append(n / step)
            n = n + hist[i+b]
    gd.numpy.take(lut, a, out=a)
    return a
```

Loading the NDVI

Next, we'll load the output of our NDVI script back into a NumPy array. We'll also define the name of our output image as `ndvi_color.tif` and create a zero-filled multi-dimensional array as a placeholder for the red, green, and blue bands of the colorized NDVI image, as you can see here:

```
# NDVI output from ndvi script
source = "ndvi.tif"

# Target file name for classified
# image image
target = "ndvi_color.tif"

# Load the image into an array
ndvi = gd.LoadFile(source).astype(gd.numpy.uint8)
```

Preparing the NDVI

We need to perform a histogram stretch on the NDVI to ensure that the image covers the range of classes that will give a meaning to the final product. We'll also create a blank array of the same shape as the image to *paint* with our class values, as you can see here:

```
# Perform a histogram stretch so we are able to
# use all of the classes
ndvi = stretch(ndvi)

# Create a blank 3-band image the same size as the ndvi
rgb = gd.numpy.zeros((3, len(ndvi), len(ndvi[0])), gd.numpy.uint8)
```

Creating classes

In this part, we set up the ranges for our NDVI classes that are broken up across a range from 0 to 255. We'll use seven classes. You can change the number of classes by adding or removing values from the "classes" list. Next, we create a **look-up table** or **LUT** to assign colors for each class. The number of colors must match the number of classes. The colors are defined as RGB values. The start variable defines the beginning of the first class. In this case, 0 is a NODATA value, which we designated in the previous script, so begin the class at 1. We then loop through the classes and extract the ranges; then we use the color assignments to add the RGB value to our placeholder array. Finally, we save the colorized image as a GeoTIFF, as you can see here:

```
# Class list with ndvi upper range values.
# Note the lower and upper values are listed on the ends
```

```
classes = [58, 73, 110, 147, 184, 220, 255]

# Color look-up table (lut)
# The lut must match the number of classes
# Specified as R, G, B tuples from dark brown to dark green
lut = [[120, 69, 25], [255, 178, 74], [255, 237, 166], [173, 232, 94],
       [135, 181, 64], [3, 156, 0], [1, 100, 0]]

# Starting value of the first class
start = 1

# For each class value range, grab values within range,
# then filter values through the mask.
for i in range(len(classes)):
    mask = gd.numpy.logical_and(start <= ndvi,
                                ndvi <= classes[i])
    for j in range(len(lut[i])):
        rgb[j] = gd.numpy.choose(mask, (rgb[j], lut[i][j]))
    start = classes[i]+1

# Save a geotiff image of the colorized ndvi.
output = gd.SaveArray(rgb, target, format="GTiff", prototype=source)
output = None
```

Here is the image we output. This is our final product for this example. Farmers can use this data to determine how to effectively irrigate and spray chemicals such as fertilizers and pesticides in a targeted, more effective, and more environment-friendly way. In fact, these classes can even be turned into a vector shapefile, which is then loaded into a GPS-driven computer on a field sprayer, which automatically applies the right amount of chemicals in the right place as a sprayer is driven around the field or in some cases even flown over the field in an airplane with a sprayer attachment.

You need to notice that even though we clipped the data to the field, the image is still a square. The black areas are the NODATA values that are converted to black. In the display software, you can make the NODATA color transparent without affecting the rest of the image, as you can see here:

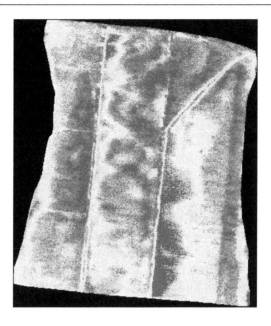

Though we created a a classified NDVI, which is a very specific type of product, the framework of this script can be altered to implement many remote sensing analysis algorithm. There are different types of NDVIs, but with relatively minor changes, you can turn this script into a tool to look for harmful algae blooms in the ocean or smoke in the middle of a forest indicating a forest fire.

This book attempts to limit the use of GDAL as much as possible to focus on what can be accomplished with pure Python and tools easily installed from PyPI. However, it is helpful to remember that there is a wealth of information on using GDAL and its associated utilities to do similar tasks. For another tutorial on clipping a raster with GDAL via its command-line utilities refer the following link:

`http://linfiniti.com/2009/09/clipping-rasters-with-gdal-using-polygons/`

Creating a flood inundation model

In the next example, we'll enter the world of hydrology. Flooding is one of the most common and devastating natural disasters, which affects nearly every population on the globe. Geospatial models are a powerful tool in estimating the impact of a flood and mitigating that impact before it happens. We often hear in the news that a river is reaching flood stage. But that information is meaningless if we can't understand the actual impact. Hydrological flood models are expensive to develop and can be very complex. These models are essential for engineers building flood control systems. However, first responders and potential flood victims are only interested to know about the impact of an impending flood.

We can understand the flooding impact in an area using a very simple and easy-to-comprehend tool called a **flood inundation model**. This model starts with a single point and floods an area with the maximum volume of water that a flood basin can hold at a particular flood stage. Usually this analysis is the worst-case scenario. Hundreds of other factors go into calculating how much water will enter into a basin from a river topping flood stage. But, we can still learn a lot from this simple first-order model.

The following image is a **Digital Elevation Model** (**DEM**) with a source point displayed as a yellow star near Houston, Texas. In real-world analysis, this point would likely be a stream gauge where you would have data about the river's water level.

As mentioned in the *Elevation data* section in *Chapter 1, Learning Geospatial Analysis with Python*, the **Shuttle Radar Topography Mission** (**SRTM**) dataset provides a nearly global DEM that you can use for these types of models. More information on SRTM data can be found at the following link:

`http://www2.jpl.nasa.gov/srtm/`

You can download the ASCII Grid data in EPSG:4326 and a shapefile containing the point as a ZIP file from the following URL:

`http://git.io/v3fSg`

The given shapefile image here is just for reference and has no role in this model:

The algorithm we are introducing in this example is called a **flood fill algorithm**, which is not really surprising. This algorithm is well known in the field of Computer Science and is used in the classic computer game *Mine Sweeper* to clear empty squares on the board when a user clicks on a square. It is also the method used for the well-known **paint bucket** tool in graphics programs such as **Adobe Photoshop**, where the paint bucket is used to fill an area of adjacent pixels of the same color with a different color. There are many ways to implement this algorithm. One of the oldest and most common ways is to **recursively** crawl through each pixel of the image. The problem with **recursion** is that you end up processing pixels more than once and then creating an unnecessary amount of work. The resource usage for a recursive flood fill can easily crash a program on even a moderately-sized image.

This script uses a four-way **queue-based** flood fill that may visit a cell more than once but ensures we only process a cell once. The queue contains unique, unprocessed cells by using Python's built-in `set` type, which only holds unique values. We use two sets called *fill* that contain the cells we need to fill, and *filled* that contain processed cells.

This example executes the following steps:

1. Extract the header information from the ASCII DEM.
2. Open the DEM as a `numpy` array.
3. Define our starting point as row and column in the array.
4. Declare a flood elevation value.
5. Filter the terrain to only the desired elevation value and below.
6. Process the filtered array.
7. Create a (1, 0) array (that is, binary array) with flooded pixels as 1.
8. Save the flood inundation array as an ASCII Grid.

Note that because this example can take a minute or two to run on a slower machine, we'll use `print` statements throughout the script as a simple way to track progress. Once again, we'll break this script up with explanations for clarity.

The flood fill function

We use ASCII Grids in this example, which means the engine for this model is completely in NumPy. We start off by defining the `floodFill()` function, which is the heart and soul of this model. The following Wikipedia article on flood fill algorithms provides an excellent overview of the different approaches:

http://en.wikipedia.org/wiki/Flood_fill

Flood fill algorithms start at a given cell and then check the neighboring cells for similarity. The similarity factor might be color or, in our case, elevation. If the neighboring cell is of the same or lower elevation as the current cell, then that cell is marked for checks of its neighbor until the entire grid is checked. NumPy isn't designed to crawl over an array in this way, but it is still efficient in handling multi-dimensional arrays overall. We step through each cell and check its neighbors to the north, south, east, and west. Any of those cells which can be flooded are added to the filled set and their neighbors added to the fill set to be checked by the algorithm.

As mentioned earlier, if you try to add the same value to a set twice, it just ignores the duplicate entry and maintains a unique list. By using sets in an array, we efficiently check a cell only once because the fill set contains unique cells, as shown in the following script:

```
import numpy as np
from linecache import getline


def floodFill(c, r, mask):
    """
    Crawls a mask array containing
    only 1 and 0 values from the
    starting point (c=column,
    r=row - a.k.a. x, y) and returns
    an array with all 1 values
    connected to the starting cell.
    This algorithm performs a 4-way
    check non-recursively.
    """
    # cells already filled
    filled = set()
    # cells to fill
    fill = set()
    fill.add((c, r))
    width = mask.shape[1]-1
    height = mask.shape[0]-1
    # Our output inundation array
    flood = np.zeros_like(mask, dtype=np.int8)
    # Loop through and modify the cells which
    # need to be checked.
    while fill:
        # Grab a cell
        x, y = fill.pop()
        if y == height or x == width or x < 0 or y < 0:
            # Don't fill
            continue
        if mask[y][x] == 1:
            # Do fill
            flood[y][x] = 1
            filled.add((x, y))
            # Check neighbors for 1 values
            west = (x-1, y)
            east = (x+1, y)
```

```
                north = (x, y-1)
                south = (x, y+1)
                if west not in filled:
                    fill.add(west)
                if east not in filled:
                    fill.add(east)
                if north not in filled:
                    fill.add(north)
                if south not in filled:
                    fill.add(south)
    return flood
```

Making a flood

In the remainder of the script, we load our terrain data from an ASCII Grid and define our output grid file name, and then we execute the algorithm on the terrain data. The seed of the flood-fill algorithm is an arbitrary point as sx and sy within the lower elevation areas. In a real-world application, these points would likely be a known location such as a stream gauge or a breach in a dam. In the final step, we save the output grid, as shown here:

```
source = "terrain.asc"
target = "flood.asc"

print("Opening image...")
img = np.loadtxt(source, skiprows=6)
print("Image opened")

# Mask elevations lower than 70 meters.
wet = np.where(img < 70, 1, 0)
print("Image masked")

# Parse the header using a loop and
# the built-in linecache module
hdr = [getline(source, i) for i in range(1, 7)]
values = [float(h.split(" ")[-1].strip()) for h in hdr]
cols, rows, lx, ly, cell, nd = values
xres = cell
yres = cell * -1

# Starting point for the
# flood inundation in pixel coordinates
sx = 2582
```

```
sy = 2057

print("Beginning flood fill")
fld = floodFill(sx, sy, wet)
print("Finished flood fill")

header = ""
for i in range(6):
    header += hdr[i]

print("Saving grid")
# Open the output file, add the hdr, save the array
with open(target, "wb") as f:
    f.write(bytes(header, 'UTF-8'))
    np.savetxt(f, fld, fmt="%1i")
print("Done!")
```

The image in the following screenshot shows the flood inundation output over a classified version of the DEM with lower elevation values in brown, mid-range values in green, and higher values in gray and white. The flood raster , which includes all areas less than 70 meters, is colored blue. This image was created with QGIS but could be displayed in ArcGIS as EPSG:4326. You could also use GDAL to save the flood raster grid as an 8-bit TIFF or JPEG just like the NDVI example to view it in a standard graphics program:

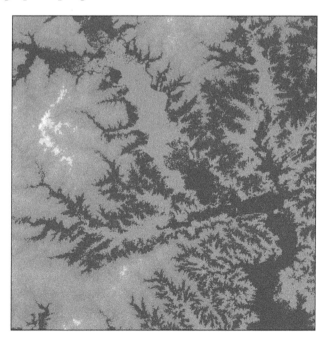

The image in the following screenshot is nearly identical except for the filtered mask from which the inundation was derived; this is displayed in yellow by generating a file for the array called `wet` instead of `fld`, to show the non-contiguous regions that were not included as part of a flood. These areas are not connected to the source point, so are unlikely to be reached during a flood event:

By changing the elevation value, you can create additional flood inundation rasters. We started with an elevation of 70 meters. If we increase that value to 90, we can expand the flood. The following screenshot shows a flood event at both 70 and 90 meters. The 90-meter inundation is the lighter blue polygon. You can take bigger or smaller steps and show different impacts as different layers:

This model is an excellent and useful visualization. However, you could take this analysis even further by using GDAL's polygonize() method on the flood mask, as we did with the island in the *Extracting features from images* section in *Chapter 6, Python and Remote Sensing*. This operation will give you a vector flood polygon. Then you could use the principles we discussed in the *Performing selections* section in *Chapter 5, Python and Geographic Information Systems*, to select buildings using the polygon to determine the population impact. You could also combine that flood polygon with the dot-density example in *Dot density calculations* section in *Chapter 5, Python and Geographic Information Systems*, to assess potential population impact of a flood. The possibilities are endless.

Creating a color hillshade

In this example, we'll combine the previous techniques in order to combine our terrain hillshade from *Chapter 7, Python and Elevation Data,* with the color classification that we used on the LIDAR. For this example, we'll need the ASCII Grid DEMs named dem.asc and relief.asc that we used in the previous chapter. We'll create a colorized DEM and a hillshade and then use PIL in order to blend them together for an enhanced elevation visualization. The code comments will guide you through the example as many of these steps are already familiar to you:

```python
import gdal_array as gd
try:
    import Image
except:
    from PIL import Image

relief = "relief.asc"
dem = "dem.asc"
target = "hillshade.tif"

# Load the relief as the background image
bg = gd.numpy.loadtxt(relief, skiprows=6)

# Load the DEM into a numpy array as the foreground image
fg = gd.numpy.loadtxt(dem, skiprows=6)[:-2, :-2]

# Create a blank 3-band image to colorize the DEM
rgb = gd.numpy.zeros((3, len(fg), len(fg[0])), gd.numpy.uint8)

# Class list with DEM upper elevation range values.
classes = [356, 649, 942, 1235, 1528,
           1821, 2114, 2300, 2700]

# Color look-up table (lut)
# The lut must match the number of classes.
# Specified as R, G, B tuples
lut = [[63, 159, 152], [96, 235, 155], [100, 246, 174],
       [248, 251, 155], [246, 190, 39], [242, 155, 39],
       [165, 84, 26], [236, 119, 83], [203, 203, 203]]

# Starting elevation value of the first class
start = 1

# Process all classes.
```

```
for i in range(len(classes)):
    mask = gd.numpy.logical_and(start <= fg,
                                fg <= classes[i])
    for j in range(len(lut[i])):
        rgb[j] = gd.numpy.choose(mask, (rgb[j], lut[i][j]))
    start = classes[i]+1

# Convert the shaded relief to a PIL image
im1 = Image.fromarray(bg).convert('RGB')

# Convert the colorized DEM to a PIL image.
# We must transpose it from the Numpy row, col order
# to the PIL col, row order (width, height).
im2 = Image.fromarray(rgb.transpose(1, 2, 0)).convert('RGB')

# Blend the two images with a 40% alpha
hillshade = Image.blend(im1, im2, .4)

# Save the hillshade
hillshade.save(target)
```

The following image shows the output that makes a great backdrop for GIS maps:

Least cost path analysis

Calculating driving directions is the most commonly used geospatial function in the world. Typically, these algorithms calculate the shortest path between points A and B, or they may take into account the speed limit of the road or even current traffic conditions to choose a route by drive time.

But what if your job is to build a new road? Or what if you are in charge of deciding where to run power transmission lines or water lines across a remote area? In a terrain-based setting, the shortest path might be to cross a difficult mountain or run through a lake. In this case, we need to account for obstacles and avoid them if possible. However, if avoiding a minor obstacle takes us too far out of our way, the cost of implementing that route may be more expensive than just going over a mountain.

This type of advanced analysis is called **Least cost path** analysis. We search an area for the route that is the best compromise of distance versus the cost of following the route. The algorithm we use for this process is called the **A-star** or **A*** algorithm. The oldest routing method is called the **Dijkstra's algorithm**, which calculates the shortest path in a network such as a road network. The A* method can do that as well, but it is also better suited for traversing a grid like a DEM. You can find out more about these algorithms on the following web pages:

- Dijkstra's algorithm: `https://en.wikipedia.org/wiki/Dijkstra%27s_algorithm`

- A* algorithm: `http://en.wikipedia.org/wiki/A-star_algorithm`

This example is the most complex in this chapter. To better understand it, we have a simple version of the program, which is text-based and operates on a 5x5 grid with randomly generated values. You can actually see how this program follows the algorithm before trying it on an elevation grid with thousands of values.

This program executes the following steps:

1. Create a simple grid with randomly-generated pseudo-elevation values between 1 and 16.

2. Define a start location in the lower-left corner of the grid.

3. Define the end point as the upper-right corner of the grid.

4. Create a cost grid that has the elevation of each cell in addition to the cell's distance to the finish.

5. Examine each neighboring cell from the start and choose the one with the lowest cost.

6. Repeat the evaluation using the chosen cell until we get to the end.

7. Return the set of chosen cells as the least-cost path.

Setting up the test grid

You simply run this program from the command line and view its output. The first section of this script sets up our artificial terrain grid as a randomly-generated NumPy array with notional elevation values between 1 and 16. We also create a distance grid, which calculates the distance between each cell to the destination cell. This value is the cost of each cell, as you can see here:

```
import numpy as np

# Width and height
# of grids
w = 5
h = 5

# Start location:
# Lower left of grid
start = (h-1, 0)

# End location:
# Top right of grid
dx = w-1
dy = 0

# Blank grid
blank = np.zeros((w, h))

# Distance grid
dist = np.zeros(blank.shape, dtype=np.int8)

# Calculate distance for all cells
for y, x in np.ndindex(blank.shape):
    dist[y][x] = abs((dx-x)+(dy-y))

# "Terrain" is a random value between 1-16.
# Add to the distance grid to calculate
# The cost of moving to a cell
cost = np.random.randint(1, 16, (w, h)) + dist

print("COST GRID (Value + Distance)")
print(cost)
print()
```

The simple A* algorithm

The A* search algorithm implemented here crawls the grid in a fashion similar to our flood fill algorithm in the previous example. Once again we use sets to avoid using recursion and to avoid duplication of cell checks. But this time, instead of checking elevation, we check the distance cost of routing through a cell in question. If the move raises the cost of getting to the end, we go with a lower-cost option, as you can see in the following script:

```
# Our A* search algorithm
def astar(start, end, h, g):
    closed_set = set()
    open_set = set()
    path = set()
    open_set.add(start)
    while open_set:
        cur = open_set.pop()
        if cur == end:
            return path
        closed_set.add(cur)
        path.add(cur)
        options = []
        y1 = cur[0]
        x1 = cur[1]
        if y1 > 0:
            options.append((y1-1, x1))
        if y1 < h.shape[0]-1:
            options.append((y1+1, x1))
        if x1 > 0:
            options.append((y1, x1-1))
        if x1 < h.shape[1]-1:
            options.append((y1, x1+1))
        if end in options:
            return path
        best = options[0]
        cset.add(options[0])
        for i in range(1, len(options)):
            option = options[i]
            if option in closed_set:
                continue
            elif h[option] <= h[best]:
                best = option
                closed_set.add(option)
            elif g[option] < g[best]:
```

```
                    best = option
                    closed_set.add(option)
                else:
                    closed_set.add(option)
            print(best, ", ", h[best], ", ", g[best])
            open_set.add(best)
        return []
```

Generating the test path

Finally, we output the least cost path on a grid. The path is represented by `1` values and all other cells as `0` values, as shown in the following lines of command:

```
print("(Y, X), HEURISTIC, DISTANCE")

# Find the path
path = astar(start, (dy, dx), cost, dist)
print()

# Create and populate the path grid
path_grid = np.zeros(cost.shape, dtype=np.uint8)
for y, x in path:
    path_grid[y][x] = 1
path_grid[dy][dx] = 1

print("PATH GRID: 1=path")
print(path_grid)
```

Viewing the test output

When you run this program, you get the following sample output (however your's will be different as it is constructed randomly):

```
COST GRID (Value + Distance)
[[13 10  5 15  9]
 [15 13 16  5 16]
 [17  8  9  9 17]
 [ 4  1 11  6 12]
 [ 2  7  7 11  8]]


(Y,X), HEURISTIC, DISTANCE
(3, 0) , 4 , 1
```

```
(3, 1) ,   1 ,   0
(2, 1) ,   8 ,   1
(2, 2) ,   9 ,   0
(2, 3) ,   9 ,   1
(1, 3) ,   5 ,   0
(0, 3) ,  15 ,   1


PATH GRID: 1=path
[[0 0 0 1 1]
 [0 0 0 1 0]
 [0 1 1 1 0]
 [1 1 0 0 0]
 [1 0 0 0 0]]
```

The grid is small enough such that you can easily trace the algorithm's steps manually. This implementation uses **Manhattan distance**, which means the distance does not use diagonal lines but only uses left, right, up, and down measurements. The search also does not move diagonally to keep things simple.

The real-world example

Now that we have a basic understanding of the A* algorithm, let's move to a more complex example. We'll use the same DEM located near Vancouver, British Columbia Canada that we used in *Creating a shaded relief* section of *Chapter 7, Python and Elevation Data*, for the relief example. The spatial reference for this grid is EPSG:26910 - NAD83 / UTM zone 10N. You can download the DEM, relief, as well as start and end points of the shapefile as a zipped package from the following URL:

http://git.io/v3fpL

We'll actually use the shaded relief for visualization. Our goal in this exercise will be to move from the start to the finish point in the lowest-cost way possible.

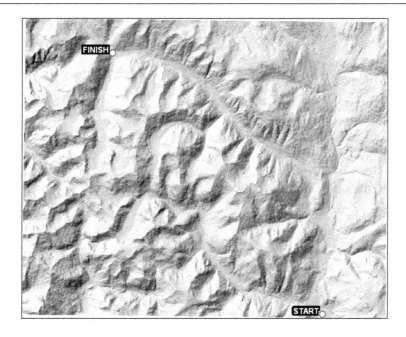

Just looking at the terrain, you can find that there are two paths that follow low-elevation routes without much change in direction. These two routes are illustrated in the following screenshot:

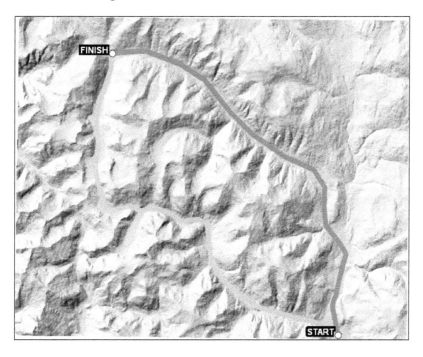

So we would expect that when we used the A* algorithm, it would be close. Keep in mind that the algorithm is only looking in the immediate vicinity, so it can't look at the whole image like we can and make adjustments early in the route based on a known obstacle ahead.

We will expand this implementation from our simple example and use Euclidian Distance or *as the crow flies* measurements; we'll also allow the search to look in eight directions instead of four. We will prioritize terrain as the primary decision point. We will use distance, both to the finish and from the start, as lower priorities to make sure we are moving forward towards the goal and not getting too far off track. Other than those differences, the steps are identical to the simple example. The output will be a raster with the path values set to 1 and the other values set to 0.

Loading the grid

The script starts out simply enough. We load the grid into a NumPy array from an ASCII Grid. We name our output path grid and then we define the starting cell and end cell, as you can see here:

```
import numpy as np
import math
from linecache import getline

# Our terrain data
source = "dem.asc"

# Output file name
# for the path raster
target = "path.asc"

print("Opening %s..." % source)
cost = np.loadtxt(source, skiprows=6)
print("Opened %s." % source)

# Parse the header
hdr = [getline(source, i) for i in range(1, 7)]
values = [float(ln.split(" ")[-1].strip()) for ln in hdr]
cols, rows, lx, ly, cell, nd = values

# Starting column, row
sx = 1006
sy = 954

# Ending column, row
dx = 303
dy = 109
```

Defining the helper functions

We need three functions to route over terrain. One is the A* algorithm and the other two assist the algorithm in choosing the next step. We'll briefly discuss these helper functions. First, we have a simple Euclidian Distance function named `e_dist`, which returns the straight-line distance between two points as map units. Next, we have an important function called `weighted_score`, which returns a score for a neighboring cell based on the elevation change between the neighbor and the current cell as well as the distance to the destination.

This function is better than distance or elevation alone because it reduces the chance of there being a tie between two cells, thus making it easier to avoid back-tracking. This scoring formula is loosely based on a concept called the **Nilsson Score**, commonly used in these types of algorithms and referenced in the Wikipedia articles mentioned earlier in this chapter. What's great about this function is that it can score the neighboring cell with any values you wish. You might also use a real-time feed to look at the current weather in the neighboring cell and avoid cells with rain or snow, as shown here:

```python
def e_dist(p1, p2):
    """
    Takes two points and returns
    the euclidian distance
    """
    x1, y1 = p1
    x2, y2 = p2
    distance = math.sqrt((x1-x2)**2+(y1-y2)**2)
    return int(distance)


def weighted_score(cur, node, h, start, end):
    """
    Provides a weighted score by comparing the
    current node with a neighboring node. Loosely
    based on on the Nisson score concept: f=g+h
    In this case, the "h" value, or "heuristic",
    is the elevation value of each node.
    """
    score = 0
    # current node elevation
    cur_h = h[cur]
    # current node distance from end
    cur_g = e_dist(cur, end)
    # current node distance from
    cur_d = e_dist(cur, start)
```

```
        # neighbor node elevation
        node_h = h[node]
        # neighbor node distance from end
        node_g = e_dist(node, end)
        # neighbor node distance from start
        node_d = e_dist(node, start)
        # Compare values with the highest
        # weight given to terrain followed
        # by progress towards the goal.
        if node_h < cur_h:
            score += cur_h-node_h
        if node_g < cur_g:
            score += 10
        if node_d > cur_d:
            score += 10
        return score
```

The real-world A* algorithm

This algorithm is more involved than the simple version in our previous example. We use sets to avoid redundancy. It also implements our more advanced scoring algorithm and checks to make sure we aren't at the end of the path before performing additional calculations. Unlike our last example, this more advanced version also checks cells in eight directions, so the path can move diagonally. There is a `print` statement at the end of this function that is commented out. You can uncomment it to watch the search crawl through the grid, as shown in the following script:

```
def astar(start, end, h):
    """
    A-Star (or A*) search algorithm.
    Moves through nodes in a network
    (or grid), scores each node's
    neighbors, and goes to the node
    with the best score until it finds
    the end.  A* is an evolved Dijkstra
    algorithm.
    """
    # Closed set of nodes to avoid
    closed_set = set()
    # Open set of nodes to evaluate
    open_set = set()
    # Output set of path nodes
    path = set()
    # Add the starting point to
```

```
# to begin processing
open_set.add(start)
while open_set:
    # Grab the next node
    cur = open_set.pop()
    # Return if we're at the end
    if cur == end:
        return path
    # Close off this node to future
    # processing
    closed_set.add(cur)
    # The current node is always
    # a path node by definition
    path.add(cur)
    # List to hold neighboring
    # nodes for processing
    options = []
    # Grab all of the neighbors
    y1 = cur[0]
    x1 = cur[1]
    if y1 > 0:
        options.append((y1-1, x1))
    if y1 < h.shape[0]-1:
        options.append((y1+1, x1))
    if x1 > 0:
        options.append((y1, x1-1))
    if x1 < h.shape[1]-1:
        options.append((y1, x1+1))
    if x1 > 0 and y1 > 0:
        options.append((y1-1, x1-1))
    if y1 < h.shape[0]-1 and x1 < h.shape[1]-1:
        options.append((y1+1, x1+1))
    if y1 < h.shape[0]-1 and x1 > 0:
        options.append((y1+1, x1-1))
    if y1 > 0 and x1 < h.shape[1]-1:
        options.append((y1-1, x1+1))
    # If the end is a neighbor, return
    if end in options:
        return path
    # Store the best known node
    best = options[0]
    # Begin scoring neighbors
    best_score = weighted_score(cur, best, h, start, end)
    # process the other 7 neighbors
```

```
            for i in range(1, len(options)):
                option = options[i]
                # Make sure the node is new
                if option in closed_set:
                    continue
                else:
                # Score the option and compare it to the best known
                    option_score = weighted_score(cur, option, h,
                        start, end)
                    if option_score > best_score:
                        best = option
                        best_score = option_score
                    else:
                        # If the node isn't better seal it off
                        closed_set.add(option)
                    # Uncomment this print statement to watch
                    # the path develop in real time:
                    # print(best, e_dist(best, end))
            # Add the best node to the open set
            open_set.add(best)
    return []
```

Generating a real-world path

Finally, we create our real-world path as a chain of ones in a grid of zeros. This raster can then be brought into an application such as QGIS and visualized over the terrain grid, as shown in the following lines of code:

```
print("Searching for path...")
p = astar((sy, sx), (dy, dx), cost)
print("Path found.")
print("Creating path grid...")
path = np.zeros(cost.shape)
print("Plotting path...")
for y, x in p:
    path[y][x] = 1
path[dy][dx] = 1
print("Path plotted.")

print("Saving %s..." % target)
header = ""
for i in range(6):
    header += hdr[i]

# Open the output file, add the hdr, save the array
```

```
with open(target, "wb") as f:
    f.write(bytes(header, 'UTF-8'))
    np.savetxt(f, path, fmt="%4i")

print("Done!")
```

Here is the output route of our search:

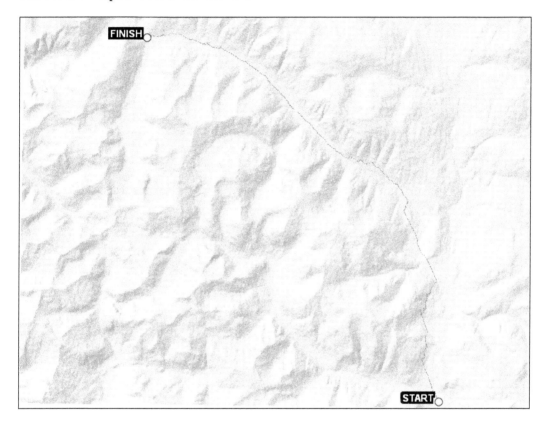

As you can see, the A* search came very close to one of our manually selected routes. In a couple of cases, the algorithm chose to tackle some terrain instead of trying to go around it. Sometimes the slight terrain is deemed less of a cost than the distance to go around it. You can see examples of that choice in this zoomed-in portion of the upper-right section of the route:

We only used two values: terrain and distance. But you could also add hundreds of factors such as soil type, water bodies, and existing roads. All of these items could serve as an impedance or an outright wall. You would just modify the scoring function in the example to account for any additional factors. Keep in mind that the more factors you add, the more difficult it is to trace what the A* implementation was *thinking* when it chose the route.

An obvious future direction for this analysis would be to create a vector version of this route as a line. The process would include mapping each cell to a point and then using the nearest neighbor analysis to order the points properly before saving as a shapefile or GeoJSON.

Routing along streets

Routing along streets uses a connected network of lines called a graph. The lines in the graph can have impedance values that discourage a routing algorithm from including them in a route. Examples of impedance values often include traffic volume, speed limit, or even distance. A key requirement for a routing graph is that all of the lines, known as edges, must be connected. Road datasets created for mapping will often have lines whose nodes do not intersect. In this example, we'll calculate the shortest route through a graph by distance. We'll use a start and end point which are not nodes in the graph, which means we'll have to first find the graph nodes closest to our start and destination.

To calculate the shortest route, we'll use a powerful pure Python graph library called **NetworkX (NX)**. NX is a general network graphing library which can create, manipulate, and analyze complex networks including geospatial networks. If pip does not install NetworkX on your system, you can find instructions for downloading and installing NX for different operating systems from this link:

`http://networkx.readthedocs.org/en/stable/`

You can download the road network as well as the start and end points , which are located along the U.S. Gulf Coast, as a ZIP file from `http://git.io/vcXFQ`.

First, we'll need to import the libraries that we're going to use. In addition to NX, we'll use the `PyShp` library to read and write shapefiles, as shown in the following script:

```
import networkx as nx
import math
from itertools import tee
import shapefile
import os
```

Next, we'll define the current directory as our output directory for the route shapefile that we'll create:

```
savedir = "."
```

Now, we'll need a function that can calculate the distance between points to populate the impedance values of our graph and to find the nodes closest to our start and destination points for the route, as seen here:

```
def haversine(n0, n1):
    x1, y1 = n0
    x2, y2 = n1
    x_dist = math.radians(x1 - x2)
    y_dist = math.radians(y1 - y2)
```

```
y1_rad = math.radians(y1)
y2_rad = math.radians(y2)
a = math.sin(y_dist/2)**2 + math.sin(x_dist/2)**2 \
    * math.cos(y1_rad) * math.cos(y2_rad)
c = 2 * math.asin(math.sqrt(a))
distance = c * 6371
return distance
```

Then we'll create another function that returns pairs of points from a list to give us line segments, which we'll use to build our graph edges:

```
def pairwise(iterable):
    """Return an iterable in tuples of two
    s -> (s0,s1), (s1,s2), (s2, s3), ..."""
    a, b = tee(iterable)
    next(b, None)
    return zip(a, b)
```

Now, we'll define our road network shapefile. This road network is a subset of a U.S. interstate highway files shapefile from the USGS, which is edited to ensure all the roads are connected:

```
shp = "road_network.shp"
```

Next, we'll create a graph with NX and add the shapefile segments as graph edges:

```
G = nx.DiGraph()
r = shapefile.Reader(shp)
for s in r.shapes():
    for p1, p2 in pairwise(s.points):
        G.add_edge(tuple(p1), tuple(p2))
```

Then we can extract the connected components as a subgraph. However, in this case we've ensured the entire graph is connected:

```
sg = list(nx.connected_component_subgraphs(G.to_undirected()))[0]
```

Next, we can read in the start and end points that we want to navigate:

```
r = shapefile.Reader("start_end")
start = r.shape(0).points[0]
end = r.shape(1).points[0]
```

Now, we loop through the graph and assign distance values to each edge using our `haversine` formula:

```
for n0, n1 in sg.edges_iter():
    dist = haversine(n0, n1)
    sg.edge[n0][n1]["dist"] = dist
```

Next, we must find the nodes in the graph closest to our start and end points to begin and end our route; we can do this by looping through all of the nodes and measuring the distance to our end points until we find the shortest distance:

```
nn_start = None
nn_end = None
start_delta = float("inf")
end_delta = float("inf")
for n in sg.nodes():
    s_dist = haversine(start, n)
    e_dist = haversine(end, n)
    if s_dist < start_delta:
        nn_start = n
        start_delta = s_dist
    if e_dist < end_delta:
        nn_end = n
        end_delta = e_dist
```

Now, we are ready to calculate the shortest distance through our road network:

```
path = nx.shortest_path(sg, source=nn_start, target=nn_end,
    weight="dist")
```

Finally, we'll add the results to the shapefile and save our route, as you can see here:

```
w = shapefile.Writer(shapefile.POLYLINE)
w.field("NAME", "C", 40)
w.line(parts=[[list(p) for p in path]])
w.record("route")
w.save(os.path.join(savedir, "route"))
```

The following screenshot shows the road network in light gray, the start and end points, and the route in black. You can see that the route cuts across the road network to reach the road nearest to the endpoint in the shortest available distance:

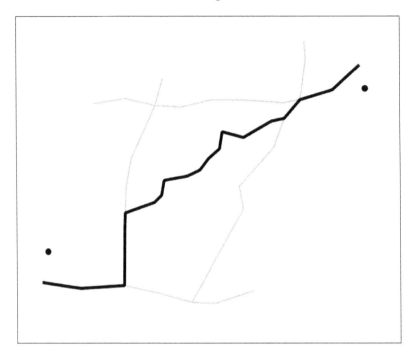

Geolocating photos

Photos taken with GPS-enabled cameras including smartphones store location information in the header of the file in a format called **exchangeable image file format** (**EXIF**) tags. These tags are based largely on the same header tags used by the **TIFF** image standard. In this example, we'll use those tags to create a shapefile with point locations for the photos and file paths to the photos as attributes.

We'll use the Python Imaging Library in this example because it has the ability to extract EXIF data. Most photos taken with smartphones are geotagged images; however, you can download the set used in this example from the following URL:

http://git.io/vczR0

First, we'll import the libraries we need including PIL for the image metadata and
PyShp for the shapefiles:

```
import glob
import os
try:
    import Image
    import ImageDraw
except:
    from PIL import Image
    from PIL.ExifTags import TAGS
import shapefile
```

Now, we'll need three functions. The first extracts the EXIF data. The second function
converts **degree, minutes, seconds (DMS)** coordinates to decimal degrees. EXIF data
stores GPS data as DMS coordinates. The third function extracts the GPS data and
performs the coordinate conversion:

```
def exif(img): # extract exif data.
    exif_data = {}
    try:
        i = Image.open(img)
        tags = i._getexif()
        for tag, value in tags.items():
            decoded = TAGS.get(tag, tag)
            exif_data[decoded] = value
    except:
        pass
    return exif_data

def dms2dd(d, m, s, i): # convert degrees, min, sec to decimal degrees

    sec = float((m * 60) + s)
    dec = float(sec / 3600)
    deg = float(d + dec)
    if i.upper() == 'W':
        deg = deg * -1
    elif i.upper() == 'S':
        deg = deg * -1
    return float(deg)

def gps(exif): # get gps data from exif
    lat = None
    lon = None
```

```
        if exif['GPSInfo']:
            # Lat
            coords = exif['GPSInfo']
            i = coords[1]
            d = coords[2][0][0]
            m = coords[2][1][0]
            s = coords[2][2][0]
            lat = dms2dd(d, m, s, j)
            # Lon
            i = coords[3]
            d = coords[4][0][0]
            m = coords[4][1][0]
            s = coords[4][2][0]
            lon = dms2dd(d, m, s, i)
    return lat, lon
```

Next, we will loop through the photos, extract the coordinates, and store the coordinates and filename in a dictionary:

```
photos = {}
photo_dir = "./photos"
files = glob.glob(os.path.join(photo_dir, "*.jpg"))
for f in files:
    e = exif(f)
    lat, lon = gps(e)
    photos[f] = [lon, lat]
```

Now, we will save the photo information as a shapefile:

```
w = shapefile.Writer(shapefile.POINT)
w.field("NAME", "C", 80)
for f, coords in photos.items():
    w.point(*coords)
    w.record(f)
w.save("photos")
```

The filenames of the photos in the shapefile are now attributes of the point locations where the photos were taken. GIS programs including QGIS and ArcGIS have tools to turn those attributes into links when you click on the photo path or the point. The following screenshot from QGIS shows one of the photos opening after clicking on the associated point using the Run feature action tool:

Summary

In this chapter, we learned how to create three very real-world products used in government, science, and industry every day. Except where this analysis is typically done with *black box* packages costing thousands of dollars, we were able to use very minimal and free cross-platform Python tools. And in addition to the examples in this chapter, you now have some more reusable functions, algorithms, and processing frameworks for other advanced analysis.

In the next chapter, we'll move into a relatively new area of geospatial analysis: real-time and near-real-time data.

9
Real-Time Data

A common saying among geospatial analysts is, *A map is outdated as soon as it's created.* This saying reflects the fact that the Earth and everything on it is constantly changing. For most of the history of geospatial analysis and through most of this book, geospatial products are relatively static. Raw datasets are typically updated from a few months to a few years. Data currency has traditionally not been the primary focus because of the time and expense needed to collect data.

Web mapping, wireless cellular modems, and low-cost GPS antennae have changed this focus. It is now logistically feasible and even quite affordable to monitor a rapidly changing object or system and broadcast these changes to millions of people online. This change is revolutionizing geospatial technology and taking it in new directions. The most direct evidence of this revolution is web mapping mash-ups using systems such as Google Maps or OpenLayers and web accessible data formats.

The term real-time data typically means near-real-time. Some tracking devices that capture real-time data may update as often as several times a second. However, the limitations of the infrastructure that broadcasts this data may constrain the output to every 10 seconds or longer. Weather radar is a perfect example. A Doppler weather radar sweeps continuously, but data is typically available online every five minutes. Given the contrast to traditional geospatial data updates, a refresh of a few minutes is real-time enough.

Web mash-ups often use real-time data. Web mash-ups are amazing and have changed the way many different industries operate. However, they are typically limited in that they usually just display preprocessed data on a map and provide developer's access to a JavaScript API. What if you want to process the data in some way? What if you want to filter it, change it, and then send it to another system? To use real-time data for geospatial analysis, you need to be able to access it as point data or a geo-referenced raster.

In this chapter, you'll learn to work with real-time geospatial data. We'll tackle the following high-level goals:

- Accessing a real-time, point location data source
- Plotting the point on a map
- Accessing a real-time raster data source
- Combining the discrete real-time data sources into more meaningful products using only Python

As with the examples in the previous chapters, the scripts are as simple as possible and designed to be read from start to finish without much mental looping. When functions are used, they are listed first, followed by the script variable declarations and finally, the main program execution.

Tracking vehicles

For our first real-time data source, we'll use the excellent NextBus API. Nextbus.com is a commercial service that tracks public transportation for municipalities, including buses, trolleys, and trains. People riding these transit lines can then track the arrival time of the NextBus. What's even better is that with the customer's permission, NextBus publishes tracking data through a **Representational State Transfer** or REST API. Using URL API calls, developers can request information about a vehicle and receive an XML document about its location. This API is a straightforward way to begin using real-time data.

If you go to Nextbus.com (`http://www.nextbus.com/#!/mu-rht/1/portlouis/`
`margstan_o`), there is a web interface to the data as well, as shown in the following
screenshot, of the city of Los Angeles, California's metro system:

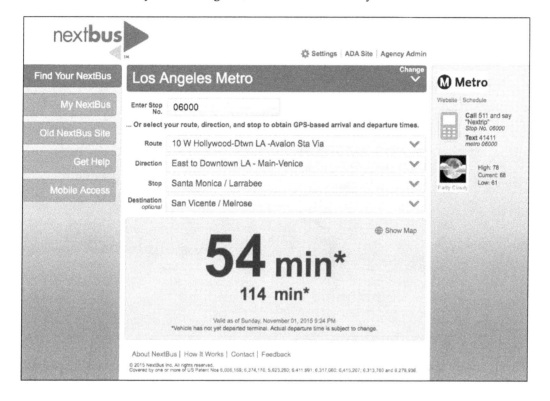

The system lets you select several parameters to learn the current location and time prediction for the next stop. On the right-hand side of the screen is a link to a Google Maps mash-up showing the transit tracking data for the particular route, as shown in the following screenshot:

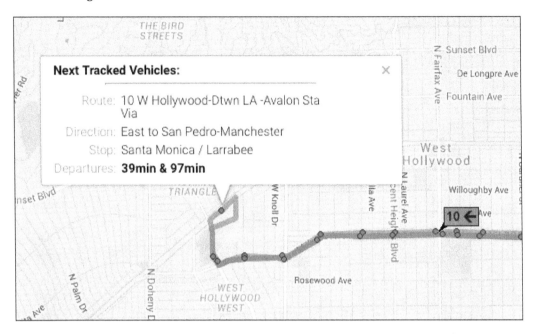

This is a very useful website but it gives us no control over how the data is displayed and used. Let's access the raw data directly using Python and the NextBus REST API to start working with real-time data.

For the examples in this chapter, we'll use the documented NextBus API that can be found at `http://www.nextbus.com/xmlFeedDocs/NextBusXMLFeed.pdf`.

The NextBus agency list

NextBus customers are called **agencies**. In our examples, we will track buses on a route for Los Angeles, California. First, we need to get some information about the agency. The NextBus API consists of a web service named publicXMLFeed in which you set a parameter named command. We'll call the `agencyList` command in a browser to get an XML document containing the agency information using the following REST URL: `http://webservices.nextbus.com/service/publicXMLFee d?command=agencyList`.

When we go to this link in a browser, it returns an XML document containing the `<agency/>` tag. The tag for Los Angeles looks as follows:

```
<agency tag="lametro" title="Los Angeles Metro"
regionTitle="California-Southern"/>
```

The NextBus route list

The `tag` attribute is the ID for Thunder Bay that we need for the other NextBus API commands. The other attributes are human readable metadata. The next piece of information that we need is the details about the Route 2 bus route. To get this information, we'll use the agency ID and the REST `routeList` command to get another XML document by pasting the URL to our web browser. Note that the agency ID is set to the `a` parameter in the REST URL: `http://webservices.nextbus.com/service/publicXMLFeed?command=routeList&a=lametro`.

When we call this URL in a browser, we get the following XML document:

```
<?xml version="1.0" encoding="utf-8" ?>
<body copyright="All data copyright Los Angeles Metro 2015.">
<route tag="2" title="2 Downtown LA - Pacific Palisades Via"/>
<route tag="4" title="4 Downtown LA - Santa Monica Via Santa"/>
<route tag="10" title="10 W Hollywood-Dtwn LA -Avalon Sta Via"/>
...
<route tag="901" title="901 Metro Orange Line"/>
<route tag="910" title="910 Metro Silver Line"/>
</body>
```

NextBus vehicle locations

So, the Mainline route ID stored in the tag attribute is simply `1` according to these results. Now, we have all the information that we need to track buses along the LA Metro route 2.

There is only one more required parameter called `t`, that is, milliseconds since the 1970 Epoch date (January 1, 1970, at midnight UTC). The epoch date is simply a computer standard used by machines to track time. The easiest thing to do in the NextBus API is to specify `0` for this value that returns data for the last 15 minutes.

There is an optional direction tag that allows you to specify a terminating bus stop in case a route has multiple buses running the route in opposite directions. If we don't specify this, the API will return the first one, which suits our needs. The REST URL to get the Mainline route for the LA Metro looks as follows: `http://webservices.nextbus.com/service/publicXMLFeed?command=vehicleLocations&a=lametro&r=2&t=0.`

Calling this REST URL in a browser returns the following XML document:

```
<?xml version="1.0" encoding="utf-8" ?>
<body copyright="All data copyright Los Angeles Metro 2015.">
<vehicle id="7582" routeTag="2" dirTag="2_758_0" lat="34.097992"
lon="-118.350365" secsSinceReport="44" predictable="true" heading="90"
speedKmHr="0"/>
<vehicle id="7583" routeTag="2" dirTag="2_779_0" lat="34.098076"
lon="-118.301399" secsSinceReport="104" predictable="true"
heading="90" speedKmHr="37"/>
 .  .  .
</body>
```

Each vehicle tag represents a location for the last 15 minutes. The last tag is the most recent location (even though XML is technically unordered).

> These public transportation systems do not run all the time. Many close down at 10:00 pm (22:00) local time. If you encounter an error in the script, use the Nextbus.com website to locate a system that is running, and change the agency and route variables to this system.

We can now write a Python script that returns the locations for a bus on a given route. If we don't specify the direction tag, NextBus returns the first one. In this example, we will poll the NextBus tracking API by calling the REST URL using the built-in Python `urllib` module demonstrated in the previous chapters. We'll parse the returned XML document using the simple built-in `minidom` module, covered in the *The minidom module* section, in *Chapter 4, Geospatial Python Toolbox*. This script simply outputs the latest latitude and longitude of the route 2 bus. You will see the agency and route variables near the top:

```
import urllib.request
import urllib.parse
import urllib.error
from xml.dom import minidom

# Nextbus API command mode
command = "vehicleLocations"
```

```python
# Nextbus customer to query
agency = "lametro"

# Bus we want to query
route = "2"

# Time in milliseconds since the

# 1970 epoch time.  All tracks
# after this time will be returned.
# 0 only returns data for the last
# 15 minutes
epoch = "0"

# Build our query url
#
# webservices base url
url = "http://webservices.nextbus.com"
# web service path
url += "/service/publicXMLFeed?"
# service command/mode
url += "command=" + command
# agency

url += "&a=" + agency
url += "&r=" + route
url += "&t=" + epoch

# Access the REST URL
feed = urllib.request.urlopen(url)

if feed:
    # Parse the xml feed
    xml = minidom.parse(feed)
    # Get the vehicle tags
    vehicles = xml.getElementsByTagName("vehicle")
    # Get the most recent one. Normally there will
    # be only one.
    if vehicles:
        bus = vehicles.pop()
        # Print the bus latitude and longitude
        att = bus.attributes
        print(att["lon"].value, ",", att["lat"].value)
    else:
        print("No vehicles found.")
```

The output of this script is simply a latitude and longitude value that implies that we now have control of the API and understand it. The output should be a coordinate value for the latitude and longitude.

Mapping NextBus locations

Now, we are ready to use this information to create our own map. The best source of freely available street mapping data is the **OpenStreetMap** (**OSM**) project: `http://www.openstreetmap.org`.

OSM also has a publicly available REST API called **StaticMapLite** to create static map images: `http://staticmap.openstreetmap.de`.

The OSM StaticMapLite API provides you with a GET API based on Google's static map API to create simple map images with a limited number of point markers and lines. A GET API, as opposed to REST, allows you to append name/value parameter pairs after a question mark on the URL. A REST API makes the parameters part of the URL path. We'll use the API to create our own NextBus API map on demand with a red pushpin icon for the bus location.

In the next example, we have condensed the previous script down to a compact function named `nextbus()`. The `nextbus()` function accepts an agency, route, command, and epoch as arguments. The command defaults to `vehicleLocations`, and the epoch defaults to `0` to get the last 15 minutes of data. In this script, we'll pass in the LA route 2 route information and use the default command that returns the most recent latitude/longitude of the bus.

We have a second function named `nextmap()` that creates a map with a blue dot on the current location of the bus each time it is called. The map is created by building a GET URL for the OSM StaticMapLite API, which centers on the location of the bus and uses a zoom level between 1-18 and map size to determine the map extent. You can access the API directly in a browser to see an example of what the `nextmap()` function does: `http://staticmap.openstreetmap.de/staticmap.php?center=40.714728,-73.998672&zoom=14&size=865x512&maptype=mapnik&markers=40.702147,-74.015794,lightblue1|40.711614,-74.012318,lightblue2|40.718217,-73.998284,lightblue3`

The `nextmap()` function accepts a NextBus agency ID, route ID, and string for the base image name of the map. The function calls the `nextbus()` function to get the latitude/longitude pair. The execution of this program loops through at timed intervals, creates a map on the first pass, and then overwrites the map on subsequent passes. The program also outputs a timestamp each time a map is saved. The `requests` variable specifies the number of passes and the `freq` variable represents the time in seconds between each loop:

```python
import urllib.request
import urllib.parse
import urllib.error
from xml.dom import minidom
import time


def nextbus(a, r, c="vehicleLocations", e=0):
    """Returns the most recent latitude and
    longitude of the selected bus line using
    the NextBus API (nbapi)
    Arguments: a=agency, r=route, c=command,
    e=epoch timestamp for start date of track,
    0 = the last 15 minutes
    """
    nbapi = "http://webservices.nextbus.com"
    nbapi += "/service/publicXMLFeed?"
    nbapi += "command={}&a={}&r={}&t={}".format(c, a, r, e)
    xml = minidom.parse(urllib.request.urlopen(nbapi))
    # If more than one vehicle, just get the first
    bus = xml.getElementsByTagName("vehicle")[0]
    if bus:
        at = bus.attributes
        return(at["lat"].value, at["lon"].value)
    else:
        return (False, False)


def nextmap(a, r, mapimg):
    """Plots a nextbus location on a map image
    and saves it to disk using the OpenStreetMap
    Static Map API (osmapi)"""
    # Fetch the latest bus location
    lat, lon = nextbus(a, r)
    if not lat:
        return False
```

```
    # Base url + service path
    osmapi = "http://staticmap.openstreetmap.de/staticmap.php?"
    # Center the map around the bus location
    osmapi += "center={},{}&".format(lat, lon)
    # Set the zoom level (between 1-18, higher=lower scale)
    osmapi += "zoom=14&"
    # Set the map image size
    osmapi += "&size=865x512&"
    # Use the mapnik rendering image
    osmapi += "maptype=mapnik&"
    # Use a red, pushpin marker to pin point the bus
    osmapi += "markers={},{},red-pushpin".format(lat, lon)
    img = urllib.request.urlopen(osmapi)
    # Save the map image
    with open("{}.png".format(mapimg), "wb") as f:
        f.write(img.read())
    return True


# Nextbus API agency and bus line variables
agency = "lametro"
route = "2"


# Name of map image to save as PNG
nextimg = "nextmap"


# Number of updates we want to make
requests = 3


# How often we want to update (seconds)
freq = 5


# Map the bus location every few seconds
for i in range(requests):
    success = nextmap(agency, route, nextimg)
    if not success:
        print("No data available.")
        continue
    print("Saved map {} at {}".format(i, time.asctime()))
    time.sleep(freq)
```

While the script runs, you'll see an output similar to the following, showing at what time the script saved each map:

```
Saved map 0 at Sun Nov  1 22:35:17 2015
Saved map 1 at Sun Nov  1 22:35:24 2015
Saved map 2 at Sun Nov  1 22:35:32 2015
```

This script saves a map image similar to the following screenshot, depending on where the bus was when you ran it:

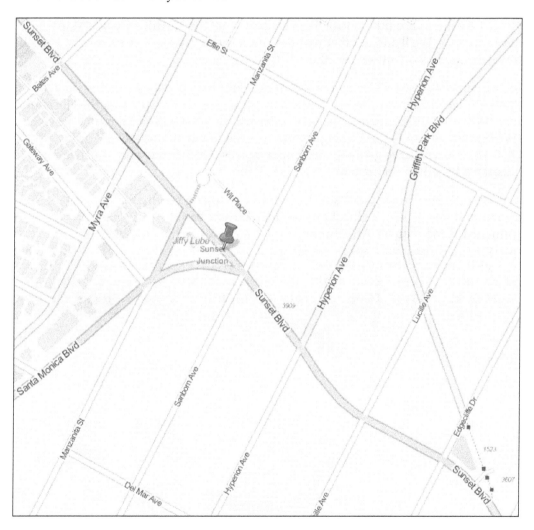

This map is an excellent example of using an API to create a custom mapping product. However, it is a very basic tracking application. To begin to develop it into a more interesting geospatial product, we need to combine it with some other real-time data source that gives us more situational awareness.

Storm chasing

So far, we created a simpler version of what the NextBus website already does. We have done it in a way that ultimately gives us complete control over the output. Now, we want to use this control to go beyond what the NextBus Google Maps mash-up does. We'll add another real-time data source that is very important to both travelers and bus operators: weather.

Iowa State University's Mesonet program provides free, polished weather data for applications. We use this data to create a real-time weather map for our bus location map. We can use the **Open Geospatial Consortium (OGC) Web Mapping Service (WMS)** standard to request a single image over our area of interest. A WMS is an OGC standard to serve georeferenced map images through the web, which are generated by a map server through an HTTP request.

The Mesonet system provides an excellent web mapping service that returns a subsetted image from a global precipitation mosaic based on a properly formatted WMS request. An example of such a request is the following query: `http://mesonet.agron.iastate.edu/cgi-bin/wms/nexrad/n0r.cgi?SERVICE=WMS&VERSION=1.1.1&REQUEST=GetMap&LAYERS=nexrad-n0r&STYLES=&SRS=EPSG:900913&BBOX=-15269659.42,2002143.61,-6103682.81,7618920.15&WIDTH=600&HEIGHT=600&FORMAT=image/png`.

As the examples in this chapter rely on real-time data, the specific requests listed may produce blank weather images if there is no activity in the area of interest. You can visit `http://radar.weather.gov/ridge/Conus/index.php` to find an area where a storm is occurring. This page contains a KML link for Google Earth or QGIS. These WMS images are transparent PNG images similar to the following sample:

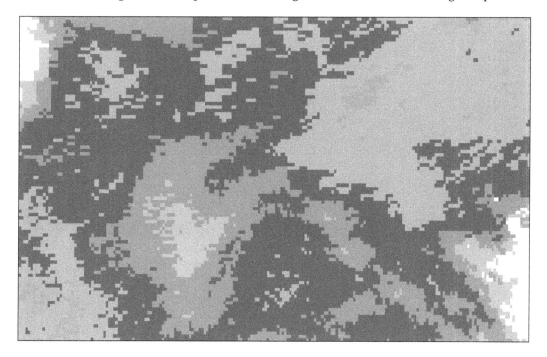

The OSM site, on the other hand, no longer provides their street maps via WMS—only as tiles. They do, however, allow other organizations to download tiles or raw data to extend the free service. The U.S. **National Oceanic and Atmospheric Administration (NOAA)** has done just this and provided a WMS interface to the OSM data, allowing requests to retrieve the single base map image that we need for our bus route:

We now have data sources to get the base map and weather data. We want to combine these images and plot the current location of the bus. Instead of a simple dot, we'll get a little more sophisticated and add the following bus icon this time:

You will need to download this icon, busicon.png, in your working directory from here: http://git.io/vlgHl.

Now, we'll combine our previous scripts and our new data sources to create a real-time weather bus map. As we are going to blend the street map and weather map, we'll need the **Python Imaging Library (PIL)** used in the previous chapters. We'll replace our nextmap() function from the previous example with a simple wms() function that can grab a map image by bounding box from any WMS service. We'll also add a function named 112m() that converts decimal degrees into meters.

The script gets the bus location, converts the location into meters, creates a two-mile (3.2 km) rectangle around the location, and then downloads a street and weather map. The map images are then blended together using PIL. PIL then shrinks the bus icon image to 30 x 30 pixels and pastes it in the center of the map, which is the bus location:

```
import sys
import urllib.request
import urllib.parse
import urllib.error
from xml.dom import minidom
import math
try:
    import Image
except:
    from PIL import Image

def nextbus(a, r, c="vehicleLocations", e=0):
    """Returns the most recent latitude and
    longitude of the selected bus line using
```

```
    the NextBus API (nbapi)"""
    nbapi = "http://webservices.nextbus.com"
    nbapi += "/service/publicXMLFeed?"
    nbapi += "command=%s&a=%s&r=%s&t=%s" % (c, a, r, e)
    xml = minidom.parse(urllib.request.urlopen(nbapi))
    # If more than one vehicle, just get the first

    bus = xml.getElementsByTagName("vehicle")[0]
    if bus:
        at = bus.attributes
        return(at["lat"].value, at["lon"].value)
    else:
        return (False, False)

def ll2m(lon, lat):
    """Lat/lon to meters"""
    x = lon * 20037508.34 / 180.0
    y = math.log(math.tan((90.0 + lat) *
                math.pi / 360.0)) / (math.pi / 180.0)
    y = y * 20037508.34 / 180
    return (x, y)




def wms(minx, miny, maxx, maxy, service, lyr, img, w, h):
    """Retrieve a wms map image from
    the specified service and saves it as a PNG."""
    wms = service
    wms += "?SERVICE=WMS&VERSION=1.1.1&REQUEST=GetMap&"
    wms += "LAYERS=%s" % lyr
    wms += "&STYLES=&"
    wms += "SRS=EPSG:900913&"
    wms += "BBOX=%s,%s,%s,%s&" % (minx, miny, maxx, maxy)
    wms += "WIDTH=%s&" % w
    wms += "HEIGHT=%s&" % h
    wms += "FORMAT=image/png"
    print(wms)
    wmsmap = urllib.request.urlopen(wms)
    with open(img + ".png", "wb") as f:
        f.write(wmsmap.read())

# Nextbus agency and route ids
agency = "roosevelt"
route = "shuttle"

# NOAA OpenStreetMap WMS service
```

```
basemap = "http://osm.woc.noaa.gov/mapcache"

# Name of the WMS street layer
streets = "osm"

# Name of the basemap image to save
mapimg = "basemap"

# OpenWeatherMap.org WMS Service
weather = "http://mesonet.agron.iastate.edu/cgi-bin/wms/nexrad/n0r.
cgi"

# If the sky is clear over New York,
# use the following url which contains
# a notional precipitation sample:
# weather = "http://git.io/vl4r1"

# WMS weather layer
weather_layer = "nexrad-n0r"

# Name of the weather image to save
skyimg = "weather"

# Name of the finished map to save
final = "next-weather"

# Transparency level for weather layer
# when we blend it with the base map.
# 0 = invisible, 1 = no transparency
opacity = .5

# Pixel width and height of the
# output map images
w = 600
h = 600

# Pixel width/height of the the

# bus marker icon
icon = 30

# Get the bus location

lat, lon = nextbus(agency, route)
if not lat:
    print("No bus data available.")
    print("Please try again later")
```

```
            sys.exit()

    # Convert strings to floats
    lat = float(lat)
    lon = float(lon)

    # Convert the degrees to Web Mercator
    # to match the NOAA OSM WMS map
    x, y = ll2m(lon, lat)

    # Create a bounding box 1600 meters
    # in each direction around the bus
    minx = x - 1600
    maxx = x + 1600
    miny = y - 1600
    maxy = y + 1600

    # Download the street map
    wms(minx, miny, maxx, maxy, basemap, streets, mapimg, w, h)

    # Download the weather map
    wms(minx, miny, maxx, maxy, weather, weather_layer, skyimg, w, h)

    # Open the basemap image in PIL
    im1 = Image.open("basemap.png")

    # Open the weather image in PIL
    im2 = Image.open("weather.png")

    # Convert the weather image mode
    # to "RGB" from an indexed PNG
    # so it matches the basemap image
    im2 = im2.convert(im1.mode)

    # Create a blended image combining
    # the base map with the weather map
    im3 = Image.blend(im1, im2, opacity)

    # Open up the bus icon image to
    # use as a location marker.
    # http://git.io/vlgHl
    im4 = Image.open("busicon.png")

    # Shrink the icon to the desired
    # size
    im4.thumbnail((icon, icon))
```

```
# Use the blended map image
# and icon sizes to place
# the icon in the center of
# the image since the map
# is centered on the bus
# location.
w, h = im3.size
w2, h2 = im4.size

# Paste the icon in the center of the image
center_width = int((w/2)-(w2/2))
center_height = int((h/2)-(h2/2))
im3.paste(im4, (center_width, center_height), im4)

# Save the finished map
im3.save(final + ".png")
```

This script will produce a map similar to the following image:

The map shows us that the bus is experiencing moderate precipitation at its current location. The color ramp, as shown in the Mesonet website screenshot earlier, ranges from light blue for light precipitation, then green, yellow, orange, to red as the rain gets heavier (or light gray to darker gray to black and white). So, at the time this map was created, the bus operator could use this image to tell their drivers to go a little slower, and bus riders would know that they might want to get an umbrella before heading to the bus stop.

As we wanted to learn the NextBus API at a low level, we used the API directly using built-in Python modules. However, several third-party Python modules exist for the API including one on PyPI simply called nextbus, which allows you to work with higher-level objects for all of the NextBus commands and provides more robust error handling that is not included in the simple examples in this chapter.

Reports from the field

In our final example in this chapter, we'll get off the bus and out into the field. Modern smart phones, tablets, and laptops allow us to update a GIS and view these updates from everywhere. We'll use HTML, GeoJSON, Leaflet JavaScript Library, and a pure Python library named Folium to create a client-server application that allows us to post geospatial information to a server, and then create an interactive web map to view these data updates.

First, we need the web form that shows your current location and updates the server when you submit the form with comments about your location. You can find the form at http://geospatialpython.github.io/Learn/fieldwork.html.

The following screenshot shows you the form:

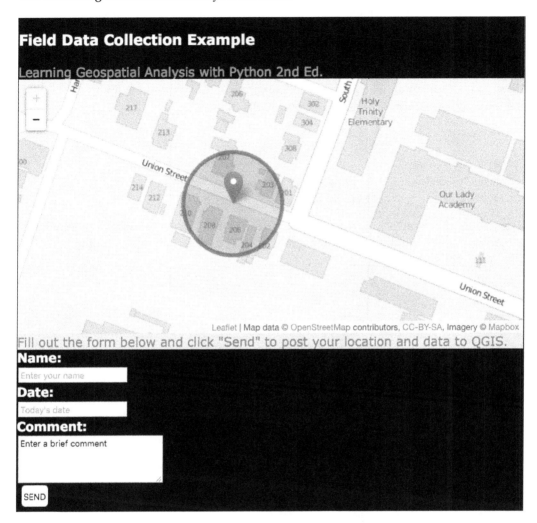

You can view the source of this form to see how it works. The mapping is done using the Leaflet library and posts GeoJSON to a unique URL on myjson.com. You can use this page on mobile devices, you can move it to any web server, or even use it on your local hard drive.

The form posts to the following URL publicly on myjson.com: `https://api.myjson.com/bins/467pm`.

You can visit this URL in a browser to see the raw GeoJSON. Next, you need to install the Folium library from PyPI. Folium provides you with a simple Python API to create Leaflet web maps. You can find more information about Folium at `https://github.com/python-visualization/folium`.

Folium makes producing a Leaflet map extremely simple. This script is just a few lines and will output a web page named `map.html`. We pass the GeoJSON URL to the map object, which will plot the locations on the map:

```
import folium

m = folium.Map()
m.geo_json(geo_path="https://api.myjson.com/bins/467pm")
m.create_map(path="map.html")
```

The resulting interactive map will display the points as markers. When you click on a marker, the information from the form is displayed. You can just open the HTML file in any browser:

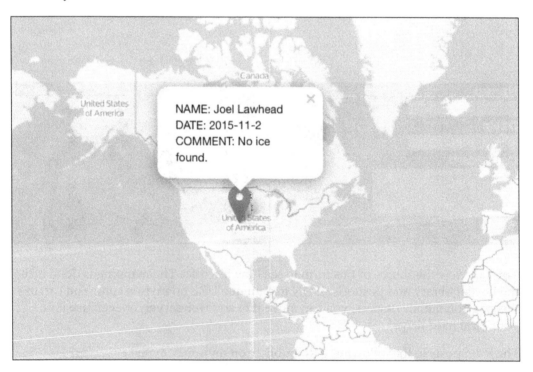

Summary

Real-time data is an exciting way to do new types of geospatial analysis, which has been made possible only recently by the advances in several different technologies, including web mapping, GPS, and wireless communications. In this chapter, you learned how to access raw feeds for real-time location data, how to acquire a subset of a real-time raster data, how to combine different types of real-time data into a custom map analysis product using only Python, and how to build client-server geospatial applications to update a GIS in real time.

As with the previous chapters, these examples contain building blocks that will let you build new types of applications using Python that go far beyond the typical popular and ubiquitous JavaScript-based mash-up.

In the final chapter, we will combine everything that you've learned so far into a complete geospatial application, which applies the algorithms and concepts to a realistic scenario.

10
Putting It All Together

In this book, we touched on all the important aspects of geospatial analysis and used a variety of different techniques in Python to analyze different types of geospatial data. In this final chapter, we will draw on nearly all of the topics that we have covered to produce one real-world product that has become very popular: a GPS route analysis report.

These reports are common to dozens of mobile app services, GPS watches, in-car navigation systems, and other GPS-based tools. A GPS typically records location, time, and elevation. From these values, we can derive a vast amount of ancillary information about what happened along the route on which that data was recorded. Fitness apps including RunKeeper.com, MapMyRun.com, Strava.com, and Nike Plus all use similar reports to present GPS-tracked exercise data from running, hiking, biking, or walking.

We will create one of these reports using Python. This program is nearly 500 lines of code, our longest yet, so we will step through it in pieces. We will combine the following techniques in this chapter:

- Accessing online data services
- Combining vector and raster data
- Parsing datasets and data feeds
- Processing raster data
- Producing graphics and reports

As we step through this program, all of the techniques used will be familiar but we will be using them in new ways.

A typical GPS report

A typical GPS report has common elements including a route map, elevation profile, and speed profile. The following screenshot is a report from a typical route logged through RunKeeper.com:

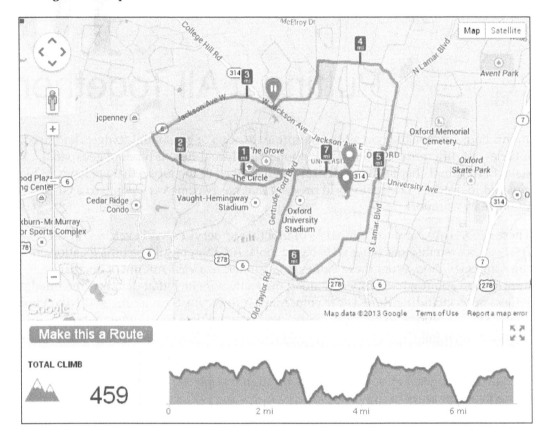

Our report will be similar, but we'll add a twist. We'll include the route map and elevation profile like this service, but we'll also add the weather conditions that occurred on this route when it was recorded.

Working with GPX-Reporter.py

The name of our program is GPX-Reporter.py. As we saw in the *Tag-based and markup-based formats* section, in *Chapter 2, Geospatial Data*, the GPX format is the most common way to store GPS route information. Nearly every program and device relying on GPS data can convert to and from GPX.

For this example, you can download a sample GPX file from at `http://git.io/vl7qi`.

You will also need to install a few Python libraries from PyPI. If you've worked through the rest of this book, you'll have most of them already:

- PIL: The Python Imaging Library
- Numpy: The multidimensional, array-processing library
- Pygooglechart: A Python wrapper for the excellent Google Chart API
- Fpdf: A simple, pure python PDF writer

Simply use `easy_install` or `pip` to install these tools. We will also be using a module called `SRTM.py`. This module is a utility to work with near-global elevation data collected during the 11-day Shuttle Radar Topography Mission in 2000 by the Space Shuttle, Endeavor. However, the official version does not yet support Python 3. You can install a Python 3 port provided by GeospatialPython.com of the library using `pip` through the following command: `pip install http://git.io/vl5Ls`.

Alternatively, you can also download the zipped file, extract it, and copy the `srtm` folder to your Python site-packages directory or your working directory: `http://git.io/vl5Ls`.

You will also need to register for a free WeatherUnderground.com developer account. This free service provides unique tools. It is the only service that provides global and historical weather data for nearly any point location: `http://www.wunderground.com/weather/api/?apiref=d7797a6597c63624`.

WeatherUnderground will provide you with a text key that you insert into a variable called `api_key` in the GPX-Reporter program before running it.

Finally, as per WeatherUnderground's terms of service, you'll need to download a logo image to be inserted into the report: `http://goo.gl/OoB4gi`.

You can review the Weather Underground Terms of Service at `http://www.wunderground.com/weather/api/d/terms.html`.

Stepping through the program

Now we're ready to work through the GPX-Reporter program. Like other scripts in this book, this program tries to minimize functions so that you can mentally trace the program better and modify it with less effort. The following list contains the major steps in the program:

- Setting up the Python logging module
- Establishing our helper functions

- Parsing the GPX data file
- Calculating the route bounding box
- Buffering the bounding box
- Converting the box to meters
- Downloading the base map
- Downloading the elevation data
- Hillshading the elevation data
- Increasing the hillshade contrast
- Blending the hillshade and base map
- Drawing the GPX track on a separate image
- Blending the track image and base map
- Drawing the start and finish points
- Saving the map image
- Calculating the route mile markers
- Building the elevation profile chart
- Getting the weather data for the route time period
- Generating the PDF report

The initial setup

The beginning of the program is to import statements followed by the Python logging module. The logging module provides a more robust way to track and log the program status than simple print statements. In this part of the program, we configure it as follows:

```
from xml.dom import minidom
import json
import urllib.request
import urllib.parse
import urllib.error
import math
import time
import logging
import numpy as np
import srtm   # Python 3 version: http://git.io/vl5Ls
import sys
from pygooglechart import SimpleLineChart
```

```
from pygooglechart import Axis
import fpdf

try:
    import Image
    import ImageFilter
    import ImageEnhance
    import ImageDraw
except:
    from PIL import Image
    from PIL import ImageFilter
    from PIL import ImageEnhance
    from PIL import ImageDraw

# Python logging module.
# Provides a more advanced way
# to track and log program progress.
# Logging level - everything at or below
# this level will output. INFO is below.
level = logging.DEBUG
# The formatter formats the log message.
# In this case we print the local time, logger name, and message
formatter = logging.Formatter("%(asctime)s - %(name)s - %(message)s")
# Establish a logging object and name it
log = logging.getLogger("GPX-Reporter")
# Configure our logger
log.setLevel(level)
# Print to the command line
console = logging.StreamHandler()
console.setLevel(level)
console.setFormatter(formatter)
log.addHandler(console)
```

This logger prints to the console, but with a few simple modifications, you can have it print to a file or even a database just by altering the configuration in this section. This module is built in Python and documented here: `https://docs.python.org/3/howto/logging.html`.

Working with utility functions

Next, we have several utility functions that are used several times throughout the program. All of these, except the functions related to time, have been used in the previous chapters in some form. The `ll2m()` function converts latitude and longitude to meters. The `world2pixel()` function converts geospatial coordinates to pixel coordinates on our output map image. Then, we have two functions named `get_utc_epoch()` and `get_local_time()` that convert the UTC time stored in the GPX file to the local time along the route. Finally, we have a haversine distance function and our simple wms function to retrieve the map images:

```
def ll2m(lat, lon):
    """Lat/lon to meters"""
    x = lon * 20037508.34 / 180.0
    y = math.log(math.tan((90.0 + lat) *
                 math.pi / 360.0)) / (math.pi / 180.0)
    y = y * 20037508.34 / 180.0
    return (x, y)


def world2pixel(x, y, w, h, bbox):
    """Converts world coordinates
    to image pixel coordinates"""
    # Bounding box of the map
    minx, miny, maxx, maxy = bbox
    # world x distance
    xdist = maxx - minx
    # world y distance
    ydist = maxy - miny
    # scaling factors for x, y
    xratio = w/xdist
    yratio = h/ydist
    # Calculate x, y pixel coordinate
    px = w - ((maxx - x) * xratio)
    py = (maxy-y) * yratio
    return int(px), int(py)


def get_utc_epoch(timestr):
    """Converts a GPX timestamp to Unix epoch seconds
    in Greenwich Mean Time to make time math easier"""
    # Get time object from ISO time string
    utctime = time.strptime(timestr, '%Y-%m-%dT%H:%M:%SZ')
    # Convert to seconds since epoch
```

```
        secs = int(time.mktime(utctime))
        return secs


    def get_local_time(timestr, utcoffset=None):
        """Converts a GPX timestamp to Unix epoch
        seconds in the local time zone"""
        secs = get_utc_epoch(timestr)
        if not utcoffset:
            # Get local timezone offset
            if time.localtime(secs).tm_isdst:
                utcoffset = time.altzone
                pass
            else:
                utcoffset = time.timezone
                pass
            pass
        return time.localtime(secs - utcoffset)


def haversine(x1, y1, x2, y2):
    """Haversine distance formula"""
    x_dist = math.radians(x1 - x2)
    y_dist = math.radians(y1 - y2)
    y1_rad = math.radians(y1)
    y2_rad = math.radians(y2)
    a = math.sin(y_dist/2)**2 + math.sin(x_dist/2)**2 \
        * math.cos(y1_rad) * math.cos(y2_rad)
    c = 2 * math.asin(math.sqrt(a))
    # Distance in miles. Just use c * 6371
    # for kilometers
    distance = c * (6371/1.609344)  # Miles
    return distance


def wms(minx, miny, maxx, maxy, service, lyr, epsg, style, img, w, h):
    """Retrieve a wms map image from
    the specified service and saves it as a JPEG."""
    wms = service
    wms += "?SERVICE=WMS&VERSION=1.1.1&REQUEST=GetMap&"
    wms += "LAYERS={}".format(lyr)
    wms += "&STYLES={}&".format(style)
    wms += "SRS=EPSG:{}&".format(epsg)
    wms += "BBOX={},{},{},{}&".format(minx, miny, maxx, maxy)
```

```
    wms += "WIDTH={}&".format(w)
    wms += "HEIGHT={}&".format(h)
    wms += "FORMAT=image/jpeg"
    wmsmap = urllib.request.urlopen(wms)
    with open(img + ".jpg", "wb") as f:
        f.write(wmsmap.read())
```

Next, we have our program variables. We will be accessing the OpenStreetMap
WMS service provided by NOAA as well as the SRTM data provided by NASA.

 We access the WMS services in this book using Python's `urllib`
library for simplicity, but if you plan to use the OGC web services
frequently, you should use the Python OWSLib package available
through PyPI at `https://pypi.python.org/pypi/OWSLib`.

We will output several intermediate products and images. These variables are
used in those steps. The `route.gpx` file is defined in this section as the `gpx` variable,
as follows:

```
# Needed for numpy conversions in hillshading
deg2rad = 3.141592653589793 / 180.0
rad2deg = 180.0 / 3.141592653589793


# Program Variables


# Name of the gpx file containing a route.
# http://git.io/vl7qi
gpx = "route.gpx"


# NOAA OpenStreetMap Basemap


# NOAA OSM WMS service
osm_WMS = "http://osm.woc.noaa.gov/mapcache"


# Name of the WMS street layer
osm_lyr = "osm"


# Name of the basemap image to save
osm_img = "basemap"


# OSM EPSG code (spatial reference system)
osm_epsg = 3857


# Optional WMS parameter
```

```
osm_style = ""

# Shaded elevation parameters
#
# Sun direction
azimuth = 315.0

# Sun angle
altitude = 45.0

# Elevation exaggerationexagerationexaggeration
z = 5.0

# Resolution
scale = 1.0

# No data value for output
no_data = 0

# Output elevation image name
elv_img = "elevation"

# RGBA color of the SRTM minimum elevation
min_clr = (255, 255, 255, 0)

# RGBA color of the SRTM maximum elevation
max_clr = (0, 0, 0, 0)

# No data color
zero_clr = (255, 255, 255, 255)

# Pixel width and height of the
# output images
w = 800
h = 800
```

Parsing the GPX

Now, we'll parse the GPX file, which is just XML, using the built-in `xml.dom.`
`minidom` module. We'll extract the latitude, longitude, elevation, and timestamps.
We'll store them in a list for later use. The timestamps are converted to `struct_time`
objects using Python's time module, which makes them easier to work with:

```
# Parse the gpx file and extract the coordinates
log.info("Parsing GPX file: {}".format(gpx))
xml = minidom.parse(gpx)
# Grab all of the "trkpt" elements
trkpts = xml.getElementsByTagName("trkpt")
# Latitude list
lats = []
# Longitude list
lons = []
# Elevation list
elvs = []
# GPX timestamp list
times = []
# Parse lat/long, elevation and times
for trkpt in trkpts:
    # Latitude
    lat = float(trkpt.attributes["lat"].value)
    # Longitude
    lon = float(trkpt.attributes["lon"].value)
    lats.append(lat)
    lons.append(lon)
    # Elevation
    elv = trkpt.childNodes[0].firstChild.nodeValue
    elv = float(elv)
    elvs.append(elv)
    # Times
    t = trkpt.childNodes[1].firstChild.nodeValue
    # Convert to local time epoch seconds
    t = get_local_time(t)
    times.append(t)
```

Getting the bounding box

We're going to need the bounding box of the route to download data from other
geospatial services. When we download data, we want the dataset to cover more
area than the route so that the map is not cropped too closely around the edges of
the route. So, we'll buffer the bounding box by 20% on each side. Finally, we'll need
the data in Eastings and Northings to work with the WMS service. Eastings and
Northings are the x and y coordinates of points in the Cartesian coordinate system in
meters. They are commonly used in the UTM coordinate system:

```
# Find Lat/Long bounding box of the route
minx = min(lons)

miny = min(lats)
maxx = max(lons)
maxy = max(lats)

# Buffer the GPX bounding box by 20%
# so the track isn't too close to

# the edge of the image.
xdist = maxx - minx
ydist = maxy - miny
x20 = xdist * .2
y20 = ydist * .2
# 10% expansion on each side
minx -= x20
miny -= y20
maxx += x20
maxy += y20

# Store the bounding box in a single
# variable to streamline function calls
bbox = [minx, miny, maxx, maxy]

# We need the bounding box in meters
# for the OSM WMS service.  We will
# download it in degrees though to
# match the SRTM file.  The WMS spec
# says the input SRS should match the
# output but this custom service just
# doesn't work that way
mminx, mminy = ll2m(miny, minx)
mmaxx, mmaxy = ll2m(maxy, maxx)
```

Downloading map and elevation images

We'll download the OSM base map first, which has streets and labels:

```
# Download the OSM basemap
log.info("Downloading basemap")
wms(mminx, mminy, mmaxx, mmaxy, osm_WMS, osm_lyr,
    osm_epsg, osm_style, osm_img, w, h)
```

This section will produce an intermediate image, as shown in the following screenshot:

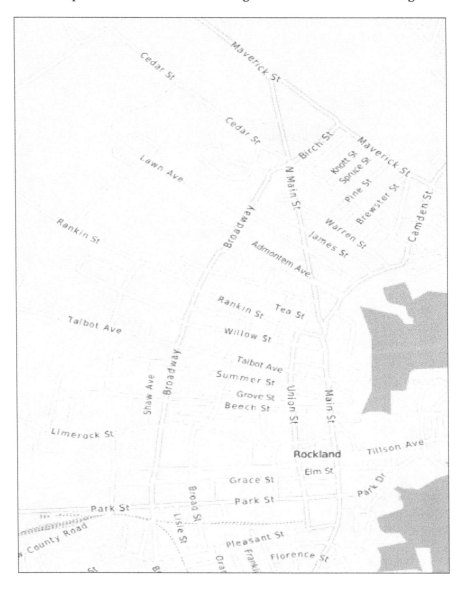

Next, we'll download some elevation data from the SRTM dataset. SRTM is nearly-global and provides a 30-90 m resolution. The Python SRTM.py module makes working with this data easy. SRTM.py downloads the datasets that it needs to make a request. So if you download data from different areas, you may need to clean out the cache located in your home directory (~/.srtm). This part of the script can also take up to 2-3 minutes to complete depending on your computer and Internet connection speeds:

```
# Download the SRTM image

# srtm.py downloader
log.info("Retrieving SRTM elevation data")
# The SRTM module will try to use a local cache
# first and if needed download it.
srt = srtm.get_data()
# Get the image and return a PIL Image object
image = srt.get_image((w, h), (miny, maxy), (minx, maxx),
    300, zero_color=zero_clr, min_color=min_clr,
      max_color=max_clr)
# Save the image
image.save(elv_img + ".jpg")
```

This portion of the script outputs an intermediate elevation image, as shown in the following screenshot:

Creating the hillshade

We can run this data through the same hillshade algorithm used in the *Creating a shaded relief* section, in *Chapter 7, Python and Elevation Data*:

```python
# Hillshade the elevation image
log.info("Hillshading elevation data")
im = Image.open(elv_img + ".jpg").convert("L")
dem = np.asarray(im)

# Set up structure for a 3x3 window to
# process the slope throughout the grid
window = []
# x, y resolutions
xres = (maxx-minx)/w
yres = (maxy-miny)/h

# Create the windows
for row in range(3):
    for col in range(3):
        window.append(dem[row:(row + dem.shape[0]-2),
                      col:(col + dem.shape[1]-2)])

# Process each cell
x = ((z * window[0] + z * window[3] + z * window[3] + z * window[6]) -
    (z * window[2] + z * window[5] + z * window[5] + z * window[8]))
\
    / (8.0 * xres * scale)

y = ((z * window[6] + z * window[7] + z * window[7] + z * window[8]) -
    (z * window[0] + z * window[1] + z * window[1] + z * window[2]))
\
    / (8.0 * yres * scale)

# Calculate slope

slope = 90.0 - np.arctan(np.sqrt(x*x + y*y)) * rad2deg

# Calculate aspect
aspect = np.arctan2(x, y)

# Calculate the shaded relief
shaded = np.sin(altitude * deg2rad) * np.sin(slope * deg2rad) \
        + np.cos(altitude * deg2rad) * np.cos(slope * deg2rad) \
        * np.cos((azimuth - 90.0) * deg2rad - aspect)

shaded = shaded * 255
```

Creating maps

We have the data that we need to begin building the map for our report. Our approach will be the following steps:

- Enhancing the elevation and base map images with filters
- Blending the images together to provide a hillshaded OSM map
- Creating a translucent layer to draw the street route
- Blending the route layer with the hillshaded map

These tasks will all be accomplished using the PIL `Image` and `ImageDraw` modules:

```
# Convert the numpy array back to an image
relief = Image.fromarray(shaded).convert("L")

# Smooth the image several times so it's not pixelated
for i in range(10):
    relief = relief.filter(ImageFilter.SMOOTH_MORE)

log.info("Creating map image")

# Increase the hillshade contrast to make
# it stand out more
e = ImageEnhance.Contrast(relief)
relief = e.enhance(2)

# Crop the image to match the SRTM image. We lose
# 2 pixels during the hillshade process
base = Image.open(osm_img + ".jpg").crop((0, 0, w-2, h-2))

# Enhance base map contrast before blending
e = ImageEnhance.Contrast(base)
base = e.enhance(1)

# Blend the the map and hillshade at 90% opacity
topo = Image.blend(relief.convert("RGB"), base, .9)

# Draw the GPX tracks
# Convert the coordinates to pixels
points = []
for x, y in zip(lons, lats):
    px, py = world2pixel(x, y, w, h, bbox)
```

```
        points.append((px, py))

    # Crop the image size values to match the map
    w -= 2
    h -= 2


    # Set up a translucent image to draw the route.
    # This technique allows us to see the streets
    # and street names under the route line.

    track = Image.new('RGBA', (w, h))


    track_draw = ImageDraw.Draw(track)


    # Route line will be red at 50% transparency (255/2=127)
    track_draw.line(points, fill=(255, 0, 0, 127), width=4)


    # Paste onto the base map using the drawing layer itself
    # as a mask.
    topo.paste(track, mask=track)


    # Now we'll draw start and end points directly on top
    # of our map - no need for transparency
    topo_draw = ImageDraw.Draw(topo)


    # Starting circle
    start_lon, start_lat = (lons[0], lats[0])
    start_x, start_y = world2pixel(start_lon, start_lat, w, h, bbox)
    start_point = [start_x-10, start_y-10, start_x+10, start_y+10]
    topo_draw.ellipse(start_point, fill="lightgreen", outline="black")
    start_marker = [start_x-4, start_y-4, start_x+4, start_y+4]
    topo_draw.ellipse(start_marker, fill="black", outline="white")


    # Ending circle
    end_lon, end_lat = (lons[-1], lats[-1])
    end_x, end_y = world2pixel(end_lon, end_lat, w, h, bbox)
    end_point = [end_x-10, end_y-10, end_x+10, end_y+10]
    topo_draw.ellipse(end_point, fill="red", outline="black")
    end_marker = [end_x-4, end_y-4, end_x+4, end_y+4]
    topo_draw.ellipse(end_marker, fill="black", outline="white")


    # Save the topo map
    topo.save("{}_topo.jpg".format(osm_img))
```

While not saved to the filesystem, the hillshaded elevation looks as follows:

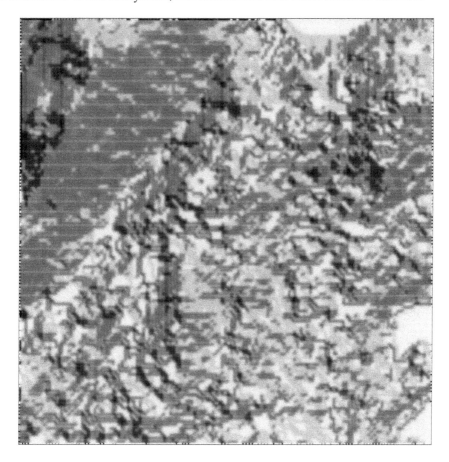

The blended topographic map looks like the following screenshot:

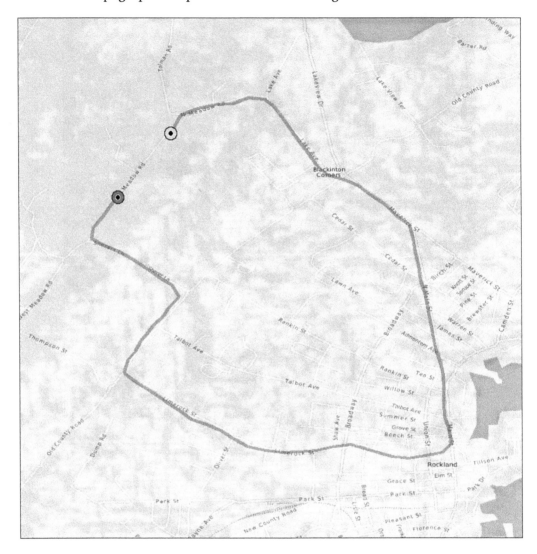

While hillshade mapping gives us an idea of the elevation, it doesn't' give us any quantitative data. To get more details, we'll create a simple elevation chart.

Measuring the elevation

Using the excellent Google Chart API, we can quickly build a nice elevation profile chart showing how the elevation changes across the route:

```
# Build the elevation chart using the Google Charts API
log.info("Creating elevation profile chart")
chart = SimpleLineChart(600, 300, y_range=[min(elvs), max(elvs)])

# API quirk -  you need 3 lines of data to color
# in the plot so we add a line at the minimum value
# twice.
chart.add_data([min(elvs)]*2)
chart.add_data(elvs)
chart.add_data([min(elvs)]*2)

# Black lines
chart.set_colours(['000000'])

# fill in the elevation area with a hex color
chart.add_fill_range('80C65A', 1, 2)

# Set up labels for the minimum elevation, halfway value, and max
value
elv_labels = int(round(min(elvs))), int(min(elvs)+((max(elvs)-
min(elvs))/2)), \
                int(round(max(elvs)))

# Assign the labels to an axis
elv_label = chart.set_axis_labels(Axis.LEFT, elv_labels)

# Label the axis
elv_text = chart.set_axis_labels(Axis.LEFT, ["FEET"])
# Place the label at 30% the distance of the line
chart.set_axis_positions(elv_text, [30])

# Calculate distances between track segments
distances = []
measurements = []
coords = list(zip(lons, lats))
for i in range(len(coords)-1):
    x1, y1 = coords[i]
    x2, y2 = coords[i+1]
```

```
        d = haversine(x1, y1, x2, y2)
        distances.append(d)
    total = sum(distances)
    distances.append(0)
    j = -1
```

Measuring the distance

In order to understand the elevation data chart, we need reference points along the *x* axis to help us determine the elevation along the route. We will calculate the mile splits along the route and place these at the appropriate location on the *x* axis of our charts:

```
# Locate the mile markers
for i in range(1, int(round(total))):
    mile = 0
    while mile < i:
        j += 1
        mile += distances[j]
    measurements.append((int(mile), j))
    j =- 1

# Set up labels for the mile points

positions = []
miles = []
for m, i in measurements:
    pos = ((i*1.0)/len(elvs)) * 100
    positions.append(pos)
    miles.append(m)

# Position the mile marker labels along the x axis
miles_label = chart.set_axis_labels(Axis.BOTTOM, miles)
chart.set_axis_positions(miles_label, positions)

# Label the x axis as "Miles"
miles_text = chart.set_axis_labels(Axis.BOTTOM, ["MILES", ])
chart.set_axis_positions(miles_text, [50, ])

# Save the chart
chart.download('{}_profile.png'.format(elv_img))
```

Our chart should now look like this:

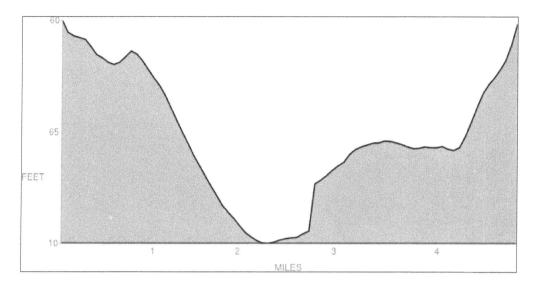

Retrieving weather data

Next, we will retrieve our final data element: the weather. As mentioned earlier, we will use the WeatherUnderground service that allows us to gather historical weather reports for any place in the world. The weather API is REST-based and JSON-based, so we'll use the urllib module to request the data and the JSON library to parse it. Note that in this section, we cache the data locally so that you can run the script offline to test, if need be. Early on in this section is where you place your WeatherUnderground API key that is flagged by the text, YOUR KEY HERE:

```
log.info("Creating weather summary")

# Get the bounding box centroid for georeferencing weather data
centx = minx + ((maxx-minx)/2)
centy = miny + ((maxy-miny)/2)

# WeatherUnderground API key
# You must register for free at wunderground.com
# to get a key to insert here.
api_key = "YOUR API KEY GOES HERE"

# Get the location id of the route using the bounding box
```

```
# centroid and the geolookup api
geolookup_req = "http://api.wunderground.com/api/{}".format(api_key)
geolookup_req += "/geolookup/q/{},{}.json".format(centy, centx)
request = urllib.request.urlopen(geolookup_req)
geolookup_data = request.read().decode("utf-8")

# Cache lookup data for testing if needed
with open("geolookup.json", "w") as f:
    f.write(geolookup_data)
js = json.loads(open("geolookup.json").read())
loc = js["location"]
route_url = loc["requesturl"]

# Grab the latest route time stamp to query weather history
t = times[-1]
history_req = "http://api.wunderground.com/api/{}".format(api_key)
name_info = [t.tm_year, t.tm_mon, t.tm_mday, route_url.split(".")[0]]
history_req += "/history_{0}{1:02d}{2:02d}/q/{3}.json" .format(*name_
info)
request = urllib.request.urlopen(history_req)
weather_data = request.read()

# Cache weather data for testing
with open("weather.json", "w") as f:
    f.write(weather_data.decode("utf-8"))

# Retrieve weather data
js = json.loads(open("weather.json").read())
history = js["history"]

# Grab the weather summary data.
# First item in a list.
daily = history["dailysummary"][0]

# Max temperature in Imperial units (Farenheit).
# Celsius would be metric: "maxtempm"
maxtemp = daily["maxtempi"]

# Minimum temperature
mintemp = daily["mintempi"]

# Maximum humidity
```

```
maxhum = daily["maxhumidity"]

# Minimum humidity
minhum = daily["minhumidity"]

# Precipitation in inches (cm = precipm)
precip = daily["precipi"]
```

Now that we have the weather data stored in variables, we can complete the final step: adding it all to a PDF report.

The `fpdf` library has no dependencies except PIL, in some cases. For our purposes, it will work quite well. We will proceed down the page and add the elements. The `fpdf.ln()` function separates rows, while `fpdf.cells` contains text and allows for more precise layouts:

```
# Simple fpdf.py library for our report.
# New pdf, portrait mode, inches, letter size
# (8.5 in. x 11 in.)
pdf = fpdf.FPDF("P", "in", "Letter")

# Add our one report page
pdf.add_page()

# Set up the title
pdf.set_font('Arial', 'B', 20)

# Cells contain text or space items horizontally
pdf.cell(6.25, 1, 'GPX Report', border=0, align="C")

# Lines space items vertically (units are in inches)
pdf.ln(h=1)
pdf.cell(1.75)

# Create a horizontal rule line
pdf.cell(4, border="T")
pdf.ln(h=0)
pdf.set_font('Arial', style='B', size=14)

# Set up the route map
pdf.cell(w=1.2, h=1, txt="Route Map", border=0, align="C")
```

```
pdf.image("{}_topo.jpg".format(osm_img), 1, 2, 4, 4)
pdf.ln(h=4.35)

# Add the elevation chart
pdf.set_font('Arial', style='B', size=14)
pdf.cell(w=1.2, h=1, txt="Elevation Profile", border=0, align="C")
pdf.image("{}_profile.png".format(elv_img), 1, 6.5, 4, 2)
pdf.ln(h=2.4)

# Write the weather summary
pdf.set_font('Arial', style='B', size=14)
pdf.cell(1.2, 1, "Weather Summary", align="C")
pdf.ln(h=.25)
pdf.set_font('Arial', style='', size=12)
pdf.cell(1.2, 1, "Min. Temp.: {}".format(mintemp), align="C")
pdf.cell(1.2, 1, "Max. Hum.: {}".format(maxhum), align="C")
pdf.ln(h=.25)
pdf.cell(1.2, 1, "Max. Temp.: {}".format(maxtemp), align="C")
pdf.cell(1.2, 1, "Precip.: {}".format(precip), align="C")
pdf.ln(h=.25)
pdf.cell(1.2, 1, "Min. Hum.: {}".format(minhum), align="C")

# Give WeatherUnderground credit for a great service
pdf.image("wundergroundLogo_black_horz.jpg", 3.5, 9, 1.75, .25)

# Save the report
log.info("Saving report pdf")
pdf.output('report.pdf', 'F')
```

You should have a PDF document in your working directory called report.pdf containing your finished product. It should look like the image shown in the following screenshot:

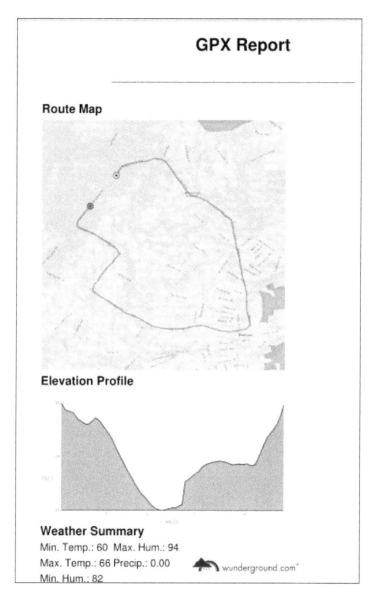

Summary

Congratulations! In this book, you pulled together the most essential tools and skills needed to be a modern geospatial analyst. Whether you use geospatial data occasionally or all the time, you will be better equipped to make the most of geospatial analysis. This book focuses on using open source tools found almost entirely within the PyPI directory for the ease of installation and integration. Even if you are using Python as a driver for a commercial GIS package or a popular library such as GDAL, the ability to test out new concepts in pure Python will always come in handy.

Index

A

B

C

Thank you for buying
Learning Geospatial Analysis with Python
Second Edition

About Packt Publishing

Packt, pronounced 'packed', published its first book, *Mastering phpMyAdmin for Effective MySQL Management*, in April 2004, and subsequently continued to specialize in publishing highly focused books on specific technologies and solutions.

Our books and publications share the experiences of your fellow IT professionals in adapting and customizing today's systems, applications, and frameworks. Our solution-based books give you the knowledge and power to customize the software and technologies you're using to get the job done. Packt books are more specific and less general than the IT books you have seen in the past. Our unique business model allows us to bring you more focused information, giving you more of what you need to know, and less of what you don't.

Packt is a modern yet unique publishing company that focuses on producing quality, cutting-edge books for communities of developers, administrators, and newbies alike. For more information, please visit our website at www.packtpub.com.

About Packt Open Source

In 2010, Packt launched two new brands, Packt Open Source and Packt Enterprise, in order to continue its focus on specialization. This book is part of the Packt Open Source brand, home to books published on software built around open source licenses, and offering information to anybody from advanced developers to budding web designers. The Open Source brand also runs Packt's Open Source Royalty Scheme, by which Packt gives a royalty to each open source project about whose software a book is sold.

Writing for Packt

We welcome all inquiries from people who are interested in authoring. Book proposals should be sent to author@packtpub.com. If your book idea is still at an early stage and you would like to discuss it first before writing a formal book proposal, then please contact us; one of our commissioning editors will get in touch with you.

We're not just looking for published authors; if you have strong technical skills but no writing experience, our experienced editors can help you develop a writing career, or simply get some additional reward for your expertise.

open source
community experience distilled

Learning Geospatial Analysis with Python

ISBN: 978-1-78328-113-8   Paperback: 364 pages

Master GIS and Remote Sensing analysis using Python with these easy to follow tutorials

1. Construct applications for GIS development by exploiting Python.

2. Focuses on built-in Python modules and libraries compatible with the Python Packaging Index distribution system – no compiling of C libraries necessary.

3. This is a practical, hands-on tutorial that teaches you all about Geospatial analysis in Python.

Python Geospatial Development

ISBN: 978-1-84951-154-4   Paperback: 508 pages

Build a complete and sophisticated mapping application from scratch using Python tools for GIS development

1. Build applications for GIS development using Python.

2. Analyze and visualize Geo-Spatial data.

3. Comprehensive coverage of key GIS concepts.

4. Recommended best practices for storing spatial data in a database.

5. Draw maps, place data points onto a map, and interact with maps.

Please check **www.PacktPub.com** for information on our titles

www.ingramcontent.com/pod-product-compliance
Lightning Source LLC
Chambersburg PA
CBHW062037050326
40690CB00016B/2967